Understanding Human Anatomy and Pathology

An Evolutionary and Developmental Guide for Medical Students

Understanding Human Anatomy and Pathology

An Evolutionary and Developmental Guide for Medical Students

Rui Diogo, Drew M. Noden, Christopher M. Smith, Julia Molnar,
Julia C. Boughner, Claudia Barrocas, Joana Bruno

CRC Press
Taylor & Francis Group
Boca Raton London New York

CRC Press is an imprint of the
Taylor & Francis Group, an **informa** business

Illustrations by Drew Noden, Christopher Smith, Julia Molnar, Joana Bruno, and Claudia Barrocas. The Authors /Illustrators Drew Noden, Christopher Smith, Julia Molnar, Joana Bruno, and Claudia Barrocas agree not to license their illustrations to any work that would compete with this Work or subsequent editions of this Work without receiving written advance permission from the Publisher.

CRC Press
Taylor & Francis Group
6000 Broken Sound Parkway NW, Suite 300
Boca Raton, FL 33487-2742

© 2016 by Taylor & Francis Group, LLC
CRC Press is an imprint of Taylor & Francis Group, an Informa business

No claim to original U.S. Government works

Printed and bound in India by Replika Press Pvt. Ltd.

Printed on acid-free paper
Version Date: 20160126

International Standard Book Number-13: 978-1-4987-5384-5 (Paperback)

Library of Congress Cataloging-in-Publication Data

Names: Diogo, Rui, author.
Title: Understanding human anatomy and pathology : an evolutionary and
developmental guide for medical students / Rui Diogo, Drew M. Noden,
Christopher M. Smith, Julia Molnar, Julia C. Boughner, Claudia Barrocas,
and Joana Bruno.
Description: Boca Raton : Taylor & Francis/CRC Press, 2016. | Includes
bibliographical references and index.
Identifiers: LCCN 2015039213 | ISBN 9781498753845 (alk. paper)
Subjects: | MESH: Anatomy. | Pathologic Processes.
Classification: LCC QM23.2 | NLM QS 4 | DDC 611--dc23
LC record available at http://lccn.loc.gov/2015039213

Visit the Taylor & Francis Web site at
http://www.taylorandfrancis.com

and the CRC Press Web site at
http://www.crcpress.com

Contents

List of Boxes

Key to boxes:

Developmental concepts	Evolutionary concepts	Pathological concepts	Learning strategies

Preface

This book is unique because it supplements existing atlases and textbooks of human anatomy with a more logical framework to learn and understand the organization of the human body. It also includes *all* the anatomical terms that students must learn in a human gross anatomy medical course. These terms, shown in bold in *Grant's Dissector*, are *all* given in bold in the main text and included in the index. The organization of this book is more versatile than most human anatomy texts in that students can skip across and refer to different sections to design their own learning plan according to their individual learning style. To wit, a student can learn the skeletal, neurovascular, and muscular head and neck as a whole using Sections 3.2 through 3.4 and then use that information to understand how all these different types of structures are associated within each anatomical region of the head and neck in Section 3.5. Alternatively, a student might prefer to first study the head and neck by region using Section 3.5, and then understand the head and neck as a whole by studying Sections 3.2 through 3.4. This in-built flexibility accounts for the deliberate overlap between the sections focused on the head and neck, to ensure the students do not miss any important structure if they choose to study only Sections 3.2 through 3.4, or alternatively only Section 3.5. As a default approach, we recommend that students study the sections in the order provided in this book.

The only major aspects that are omitted in this book are the brain, internal organs, and external sexual organs. The justification for leaving out this admittedly significant content is that the evolution and development of the musculoskeletal system and related neurovascular structures are much more studied and thus better known than the evolutionary history and ontogeny of most other structures, particularly internal organs such as the stomach or the liver, for instance. Thus, unfortunately, it is not possible at the moment to adequately provide evolutionary and developmental details about the origins and subdivisions of all these organs.

Therefore, this book should be seen as a unique, irreplaceable resource to better understand and memorize—using developmental, evolutionary, and pathological concepts—the human musculoskeletal system and associated neurovascular structures, which include most of the body structures that students have to learn in gross anatomy courses. Moreover, the anatomy of various internal organs that are deeply related to musculoskeletal and neuro-vascular structures of the head, neck, trunk, and limbs—such as the heart, ears, eyes, tongue, and nose—is covered in some detail in this book as well. In most cases, developmental, evolutionary, and pathological concepts as well as learning strategies are highlighted in boxes (yellow, green, blue, and gray boxes, respectively) to make the text easier to follow.

Finally, this book includes 88 original, high quality plates depicting concepts of embryology (pp. 223–231) and all of the major structures covered in a human gross anatomy course (pp. 232–311). These illustrations were created by trained medical and scientific illustrators who coauthored this book, collaborating with the writers to make sure that all the structures and concepts described in the text are accurately portrayed in the illustrations.

Acknowledgments

We are thankful to all our colleagues and students for all discussions, comments, and suggestions concerning any subject dealt within this book. We are particularly thankful to our students because their wise questions and immense curiosity obliged us to learn more, every year, and this continuous learning resulted in this book, which is dedicated to all of them. We also want to thank The National Evolutionary Synthesis Center (NESCent) for their financial support, through the National Science Foundation (NSF), which was crucial for the writing of this book. NESCent has made a unique, invaluable contribution to students, teachers, and researchers in the United States and also across the planet through their collaborations and exchange of ideas.

Authors

Rui Diogo is an award-winning (American Association of Anatomists and Anatomical Society of Great Britain and Ireland) assistant professor at Howard University College of Medicine and a resource faculty at the Center for the Advanced Study of Hominid Paleobiology of George Washington University. He is the author/coauthor of numerous publications, coeditor of the books *Catfishes* and *Gonorynchiformes and Ostariophysan Interrelationships*, and coauthor of *Muscular and Skeletal Anomalies in Human Trisomy in an Evo-Devo Context*. He is the sole author or first author of several monographs, including five atlases of baby gorillas and adult gorillas, chimpanzees, hylobatids, and orangutans, and the books *Morphological Evolution, Aptations, Homoplasies, Constraints and Evolutionary Trends*; *The Origin of Higher Clades: Osteology, Myology, Phylogeny and Evolution of Bony Fishes and the Rise of Tetrapods*; *Muscles of Vertebrates*; and *Comparative Anatomy and Phylogeny of Primate Muscles and Human Evolution*.

Drew M. Noden is a professor of embryology and animal development at Cornell University. His research explores the development of vertebrate craniofacial muscles, blood vessels, and skeletal tissues with the goal of understanding the interactions between precursor populations and their neighbors that may be important to the initiation and progression of cell differentiation and the morphogenesis of musculoskeletal complexes. Over the course of his long and distinguished career, Drew has authored numerous high-impact publications, received many competitive grants, and been recognized for his scientific achievement by organizations such as the American Association of Anatomists and the International Association of Dental Research.

Christopher M. Smith earned his MA from the Johns Hopkins University School of Medicine in medical and biological illustration. He combines biomedical research with traditional and digital visualization techniques. Chris has received several awards for his work, including the Vesalius Award for the best young medical illustrator within North America.

Julia Molnar is a functional anatomist and biomedical illustrator. Her research focuses on evolutionary changes in musculoskeletal anatomy and their effects on function, particularly locomotion. She has published on the evolution of locomotion in early tetrapods and crocodylomorphs and the use of virtual models in paleontology. She is also an accomplished scientific illustrator whose work has appeared in numerous books, museums, scientific journals, and popular science publications. Julia earned her PhD from the Royal Veterinary College, London and is currently a postdoctoral fellow in the anatomy department at Howard University in Washington, DC.

Julia C. Boughner is an assistant professor at the University of Saskatchewan College of Medicine. Her research background in physical anthropology informs her "evo-devo" research on vertebrate heads and teeth, and how they reveal new insights into primate developmental biology and evolution. She is the coeditor of *Developmental Approaches to Human Evolution* and author/coauthor of many publications in the fields of developmental biology and evolutionary morphology. Her laboratory's research discoveries are supported by the Natural Sciences and Engineering Research Council of Canada.

Claudia Barrocas is a graduate student in biological illustration at the University of Aveiro, Portugal. She has studied scientific illustration, and collaborates with one of the most renowned scientific illustrators in Europe, Fernando Correia, who is associated, in an indirect way, with the careful execution of this book.

Joana Bruno is a scientific illustrator and a graduate student in prehistoric archaeology. She specializes in natural science illustration, archaeological and paleontological reconstructions, and illustrated infographics. She earned her master's degree in scientific illustration at the Instituto Superior de Educação e Ciências and Universidade de Évora in Portugal, and is currently pursuing her PhD at Universitat Autònoma de Barcelona. She received various awards for her illustrations, including some of the illustrations created for this book.

Chapter 1

About the Book

Anatomy is the oldest formal discipline in medicine, and this book builds upon the exceptional contributions made by outstanding scholars. Among them, we want to pay special tribute to some of the most accurate and most frequently used atlases and dissector manuals for human gross anatomy: *Netter's Atlas of Human Anatomy* (6th edition, 2014), *Gray's Atlas of Anatomy* (2nd edition, 2014), *Grant's Atlas of Anatomy* (13th edition, 2012), and *Grant's Dissector* (15th edition, 2015). We are particularly grateful to the authors of these books, and their predecessors, for their remarkable and crucial contribution to the learning and teaching of human gross anatomy. The present book is a way for us to acknowledge our deep gratitude for the amazing work done by these authors and by the authors of similar atlases and textbooks.

To maintain clarity and narrative flow for the students, we have not included citations and parenthetical asides in this book. We are, of course, not trying to deny credit to other authors, as we just acknowledged some of these books, but we provide at the end of this book a brief, summarized list of suggested readings related to the content of each chapter. In fact, the main goal of this book is to provide medical students a unique, and much easier and comprehensive way of learning and understanding human gross anatomy by combining state-of-the-art knowledge about human anatomy, evolution, development, and pathology. However, the purpose is *not* to provide a huge amount of information that might be interesting for students, but that lies outside of the aspects they will need to learn and understand for their gross anatomy courses and their board exams in particular. The purpose is instead to add the evolutionary, pathological, and developmental information in a way that will *reduce* the difficulty of, and amount of total time spent in, learning gross anatomy, by making learning more logical and systematic. Also, as the plan is to synthesize data that would normally be available for students only by consulting several books focused on human anatomy, evolution, development, and pathology at the same time, this book was put together in such a way that it is relatively short and easily transportable. So, students can take the book anywhere and use it to learn and understand *all the structures* that they have to normally learn for any gross anatomy course, with the exception of the internal and sexual organs, as explained above.

Therefore, to describe anatomical structures, we use the brief, direct writing style of "dissector" textbooks frequently used by students, highlighting in bold *all the anatomical structures* that are normally mandatory for medical students to know, and provide an index where all these structures, as well as other key concepts/terms also highlighted in bold in the text, can be easily found. Concerning the illustrations, we carefully selected a set of anatomical illustrations that follow the style of those used in human anatomical atlases but that, in our opinion, are simpler in their overall configuration and sufficient in number to show and make it easier to better understand the anatomical structures highlighted in bold in the text, without overwhelming the students with visual information. This is particularly important because the uniqueness of this book is that it includes these high-quality anatomical illustrations *plus* a selective set of illustrations and schemes to go along with the evolutionary, developmental, and pathological parts of the text.

It is also important to emphasize that this book is also a way for us—all the authors of this book except Drew Noden—to pay tribute to the truly remarkable scientific contribution of Drew to the topics presented and discussed in the book. In fact, several parts of this book were adapted from the excellent notes and diagrams that he has meticulously prepared for the courses that he taught at Cornell University. Thank you, Drew, for this and for your amazing career, as well as your kindness.

Chapter 2

Introduction

2.1 Conserved Features of Vertebrate Embryology

Most anatomical features of human embryos (Plate 2.1) are common to all vertebrate embryos, based on a pattern that emerged over 400 million years ago. Examples include the **notochord**, a collagenous rod that extends the length of the body axis, and **somites**, segments of mesoderm arranged on either side of the notochord. The tissues that will form these structures are produced by rearrangements of embryonic cells during a stage called **gastrulation**, when the embryo is less than two weeks old. These tissues are: (1) the **ectoderm**, which forms the skin and nervous system; (2) the **mesoderm**, which forms the musculoskeletal system and circulatory systems; and (3) the **endoderm**, which forms most of the digestive and respiratory systems. Very soon after gastrulation, organs begin to form. The first structures to appear are the primordia of the brain and spinal cord, backbone and muscles, digestive tube, and cardiovascular systems (Plate 2.2). These early organ systems will expand, a few additional ones added, and by the 10–12-mm stage, all organ systems are in place and most are functional. Which of these structures, based on their location and organization, resemble those found in the adult? For example, do you have somites and a notochord? If not, why are they present in all vertebrate embryos, including humans? The reason is phylogenetic constraint: We, humans, have an evolutionary history from our ancestors, and the organizational configuration that was laid down during early vertebrate evolution has remained with few modifications. The fact that embryos continue to adhere to this very ancient configuration is due to **developmental constraint**.

Somites are clusters of mesodermal cells that aggregate tightly together in a cranial to caudal sequence on both sides of the **hindbrain** and **spinal cord**. Later, these blocks rearrange to form **vertebrae** and **ribs**, all the voluntary muscles of the body and limbs, and most axial connective tissues (see Section 6.1). Equally important, somites establish a segmental pattern that is imposed on the spinal cord and peripheral nerves that emerge from it. In the embryo, **segmentation** allows for the formation of repeated sets of nearly identical tissues, thereby simplifying the "blueprints" necessary to generate each segment. Secondarily, each segment receives signals that give it its unique spatial identity, allowing, for example, cervical and thoracic vertebrae to have different shapes. Segments also define spatial compartments that allow the nerves, muscles, and skeletal structures to contact each other and form stable relationships that persist in the adult. The clear segmentation of adult structures such as vertebrae and the peripheral nerves that emerge from the spinal cord reflects the phylogenetic and developmental constraints established by our ancestors and maintained during earlier developmental stages. Of course, many new structures have been added over the course of vertebrate evolution—limbs are a good example. As limbs evolved, the embryo co-opted nearby progenitor populations derived from somites to form appendicular muscles, which do not maintain their cranio-caudal segmented pattern. However, since the basic pattern of limb musculoskeletal organization was established, it has remained largely unchanged during early stages of development.

Plate 2.3 summarizes the key anatomical features that define the common embryonic body pattern of vertebrate embryos. These features include: a dorsal, hollow **neural tube** with regional specializations; a **notochord** located in the midline, immediately ventral to the neural tube; a series of *segmentally arranged* **somites** that later form axial musculoskeletal tissues; a ventral **gut tube** that forms the lining of organs associated with respiration and digestion; a **coelomic cavity** that later becomes subdivided into **pleural**, **pericardial**, and **peritoneal cavities**; a ventral heart tube from which blood flows through a series of segmentally arranged arteries; and a **body wall** that surrounds the embryo except at the umbilicus. The developing embryo is well protected. In mammalian embryos, the first organ system to become functional is the placenta. The placenta is composed of both maternal uterine tissues and membranes derived from embryonic tissues, which establish intimate associations with the lining of the maternal uterus. The placenta is an amazingly complex structure capable of serving functions later carried out by the neonatal liver, kidneys, lungs, endocrine, and digestive systems. While it is usually selective about what molecules are allowed to pass into the embryonic circulation, it is not impermeable. Both acute and chronic exposure to complex environmental chemicals, some viruses, and physical agents presents potential risks to the embryo, and many of these—alone or in combination—have not been rigorously assessed. Additionally, agents that disrupt maternal endocrine and metabolic functions can have severe secondary consequences for embryonic and placental development.

2.2 Causes and Mechanisms of Developmental Pathologies

While we have emphasized the common—constrained—aspects of early development, each species also has unique embryonic and placental structures and functions. Genetic and metabolic activities in the embryo are also highly specialized and stage-specific and, especially at early stages when organ systems are being established, are particularly vulnerable to disruptions. These can manifest as structural alterations, often apparent at birth, but also as more subtle metabolic changes that carry forward into and through adulthood and even subsequent generations. Genetic factors (including polymorphisms and mutations) establish limits and functional boundaries for **metabolic pathways** and **signaling cascades**. Within this range, however, environmental influences that skew maternal, placental, or embryonic/fetal function can have substantial and irreparable effects. For the purposes of understanding both normal embryology and developmental abnormalities, the extent to which extrinsic influences—anything outside of the embryo/fetus—affect the embryo or placenta must be considered.

Developmental **abnormalities** can result from disruptions of **cell proliferation**, **cell movement**, **cell differentiation**, **cell survival**, or **morphogenesis**. Abnormalities arising during prenatal stages of development are commonly called **congenital abnormalities** or **birth defects**. Some abnormalities may not show clinical signs until the affected systems become functional (e.g., sensory systems, locomotion, reproduction) or stressed (metabolic, endocrine), or they may compromise cell functions in ways that are not recognized until adulthood; for example, increasing risks of cancer or heart disease. Clinically relevant abnormalities are present in 3% of live births, or ~320 babies in the United States every day, and an equal number are subsequently detected prior to the age of one year (data available from the Center for Disease Control website). About half of all neonatal deaths are attributable to identified developmental problems, ranging from gross structural abnormalities to those that are more subtle, such as low birth weight and placental insufficiency. Defects frequently are part of a **syndrome**, a set of abnormalities that often appear together, and are assumed to result from the same genetic variation or environmental insult. Birth defects contrast with **anatomical variations**, which are seen in the karyotypically/genetically "normal" population: You will surely see such variations in the human cadavers you dissect in your gross anatomy course.

All structures and functions in embryonic and placental cells and tissues are potentially subject to disruption by outside factors, broadly categorized as **teratogens**. In the past, emphasis was placed on chemicals or physical factors (e.g., radiation, heat) that reduced viability and caused morphological abnormalities. Now the range of disrupters is much broader, and includes nutritional factors as well as the collective effects of multiple agents, each of which alone is not significantly problematic. It is estimated that *40%–70% of human embryos die and are aborted during the first four weeks of development*, mostly during the gastrulation stage when intraembryonic reorganizations, activation of several thousand genes, and initial apposition with uterine epithelium and placental formation all occur simultaneously over a period of less than two days. In one study, it was found that at least half of early aborted embryos have major genetic lesions, such as lost, broken, or extra chromosomes.

Traditional explanations of the causes of defects focus on single-gene mutations or environmental toxins. However, while these factors may be important contributors, most developmental defects cannot be adequately explained by them. We now know that an optimal outcome—a healthy organism, organs, tissues, etc.—requires cooperative and integrated spatial and temporal regulation of every ongoing process within all cells of the embryo. Each of these processes is the result of interactions involving many genes and their products with many elements of their environment and nutrition. Teratogens can target the embryo directly or primarily affect the mother or the placenta. Every cell in each of these sites continually monitors and responds to all aspects of the chemical and physical world around it, using a combination of receptors, channels, pores, and endocytotic vesicles, and then integrates the complex signals it receives through multiple interconnected metabolic and gene-regulatory pathways. These pathways determine whether the cells divide or not, survive or die, maintain the same phenotype or change, and move or remain stationary. Some extrinsic factors (e.g., radiation, some steroid analogs) act directly upon **DNA** (deoxyribo-nucleic acid, the molecule that stores hereditary information) or other cell structures, but most interfere with the metabolic and signal transduction pathways that integrate and regulate cell activities.

2.3 A Holistic Approach to Anatomy

The points mentioned above emphasize the importance of a holistic approach to developmental and gross anatomy and pathology. Understanding gross anatomy can inform pathology, and knowledge about pathology, combined with evolutionary and developmental information, can also inform—and thus make easier to understand—human gross anatomy, as we will show in this book. In fact, in the nineteenth century and the beginning of the twentieth century, many physicians and medical researchers were also comparative anatomists and/or evolutionary biologists. Rudolf Virchow (1821–1902) and Etienne Serres (1786–1868) are two particularly elucidatory examples of this tradition. The current gaps between medicine, evolutionary biology/comparative anatomy, and developmental biology appeared only later, particularly with the more gene-centered approach to which many biologists and medical researchers started to subscribe in the second half of the twentieth century. These separations have been breaking down over the last few years as molecular genetics merges evolutionary and developmental processes—(Evo-Devo) and more recently evolutionary medicine and evolutionary developmental anthropology—and blurs the distinction between "normal" and pathological mechanisms.

One of the unfortunate heritages that resulted from the decoupling of medicine—particularly medical gross anatomical courses—from evolutionary biology and comparative anatomy was the treatment of the human body and its maladies as distinct from other living systems. For this reason, physicians, students, and medical researchers did not incorporate the understanding of evolutionary changes and constraints. For instance, many textbooks and documentaries describe the human body as a "perfect machine," contrary to the principles of biological systems. Our bodies, like those of all other living organisms, may include some examples of structures wonderfully adapted to the niches we occupy, but parts will fail, due to injury or disease or aging, and some functions are irreversibly lost. The marvelous advances in health, nutrition, and treatments of the twentieth century nearly doubled the life expectancy in many (but not all!) parts of the planet. While this increase in life expectancy is cause for celebration, it has also revealed patterns of later life illnesses that were not part of our evolutionary history.

An emblematic example that shows the problems, and the profound implications, of the decoupling between medicine and evolutionary/comparative anatomy is the way physicians and medical students and researchers see the upper and lower limbs. Specifically, the human bipedal **anatomical position** used in anatomical and medical textbooks (Figure 2.1) does not correspond to the reality of our evolutionary and developmental origins. We are descended from quadrupedal animals, and this evolutionary history is also reflected in our embryonic development. Figure 2.2 shows the adult human body with the typical crawling quadrupedal posture of most adult mammals and comparative anatomical directional terms (**rostral**, i.e., toward the head, and **caudal**, i.e., toward the tail, which correspond to superior and inferior in bipeds, respectively). Using this drawing, it becomes easier to

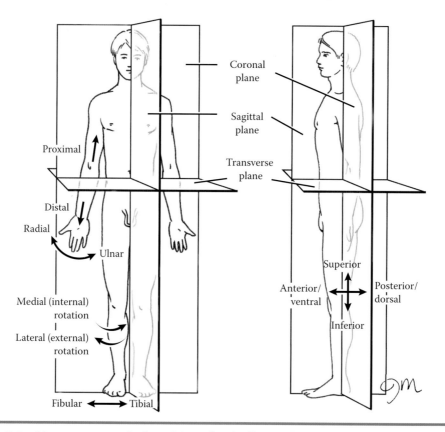

Figure 2.1 **Human anatomical position classically used in anatomical and medical textbooks. The figure stands bipedally, the forearms are supinated so that the 1st digit (thumb) is on the lateral side, and the feet point forward so that the 1st digit (big toe) is on the medial side.**

Rostral ◄─────────────► Caudal

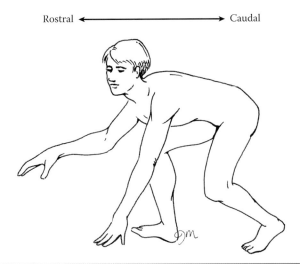

Figure 2.2   Human in a quadrupedal posture more appropriate to our evolutionary history. Both sets of limbs are held below the body with the 1st digit (thumb and big toe) on the medial side.

understand, for instance, the configuration of the nerves in the human body, particularly of the trunk, and why we use terms such as "**dorsal rami**," or "**dorsal sacral foramina**," to designate structures that are posterior according to the anatomical position given in most textbooks. The basic **body configuration** (i.e., overall configuration of the body, not to be confused with body posture) of our basal **tetrapod** ancestors that first left the water to walk on the ground with four limbs was more similar to that of modern quadrupedal amphibians such as salamanders (Figure 2.3) than that of modern quadrupedal mammals, in which the limbs are highly rotated from

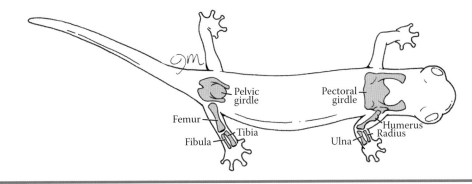

Figure 2.3   Limb anatomy of a salamander, representing the plesiomorphic tetrapod body plan, and showing correspondences between the fore- and hindlimb bones as humerus/femur, tibia/radius, and fibula/ulna.

the basal condition. In the salamander, it is easy to draw a topological correspondence (based on similar **anatomical configuration**) between individual structures of the forelimb and hindlimb (the upper and lower limb in
human anatomy): The "thumb" and radius correspond to the "big toe" and
tibia, which are the most rostral bones of the hand/forearm and of the foot/
leg, respectively, and thus the ulna corresponds to the fibula. Importantly,
this figure also shows that the dorsal (extensor) side of the femur in the
hindlimb, where muscles homologous to the human quadriceps femoris are
located, corresponds to the dorsal (extensor) side of the humerus in the forelimb, where muscles homologous to the human triceps brachii are located.

It is much simpler to understand the musculature of the human body
once you know that true (i.e., evolutionarily and developmentally) dorsal
muscles encompass the *extensor* groups of limb muscles, while true ventral muscles encompass the *flexor* groups. Each group in the upper and
lower limb is innervated by a different set of nerves, as will be explained
in Chapters 4 and 5. These functional and topological relationships account
for why all the intrinsic hand muscles—which are derived from true ventral
or palmar (flexor) muscles—are innervated by the ventral ulnar and median
nerves, and why all the intrinsic foot muscles—which are derived from true
ventral or plantar (flexor) muscles—are innervated by branches of the tibial
nerve. The main problem with the traditional anatomical position of humans
is that, while it places the true dorsal sides of the arm, forearm, and hand in
a dorsal position (usually designated posterior in human anatomy), it places
the true dorsal sides of the thigh, leg, and foot in a ventral (usually designated anterior in human anatomy) position. The anatomists who invented
the terms we use to describe the movements of the human body were probably aware of this problem, which is why they used the term "extensor" to
designate muscles on the anterior side of the leg such as the extensor digitorum longus and extensor hallucis longus, and logically designated it the
"extensor side of the leg." However, for some reason, the function of these
muscles called "extensors" was designated "dorsiflexion of the foot" instead
of, more logically, "extension of the foot"; likewise, the opposite movement
was designated "plantar flexion of the foot" rather than the more logical "flexion of the foot." Even more confusingly, thigh muscles such as the
rectus femoris that lie on the same side of the body as these leg extensors
are called flexors of the thigh. By coining this terminology, early anatomists
birthed the confusion that most often impedes medical students' understanding of the muscles of the human body and of their function. It would be
far more logical to say that all the true dorsal muscles of both the thigh and

leg—thus including the rectus femoris—are extensors, rather than calling the leg ones extensors and the thigh ones flexors.

Figure 2.4 shows a human body in which the right side (left side of the figure) is in the anatomical position followed by most textbooks and atlases. On the left side of the body (right side of the figure), the thigh, leg, and foot are rotated almost 180° to show that the so-called "posterior" sides of the thigh and leg, and "plantar" side of the foot correspond to the anterior sides of the arm and forearm, and palmar side of the hand, respectively. The fact that these sides of the upper and lower limbs do correspond to each other also becomes clear when one studies **human development**: During human development, the upper and lower limbs are rotated 90° each, in different directions, resulting in the thumb being most lateral and the big toe being most medial within their respective limbs (Figure 2.5). This correspondence

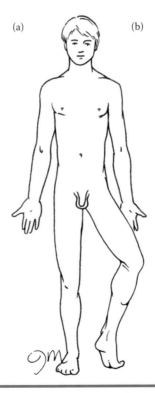

Figure 2.4 **Comparison between standard anatomical position, on the figure's right side (a), and a position with the lower limb pronated, on the figure's left side (b). In (a), the anterior view shows the flexor musculature of the upper limb and the extensor musculature of the lower limb. In (b), the lower limb is pronated almost 180° so that the flexor musculature of the leg and the plantar surface of the foot face anteriorly, corresponding to the flexor musculature of the arm and palmar surface of the hand.**

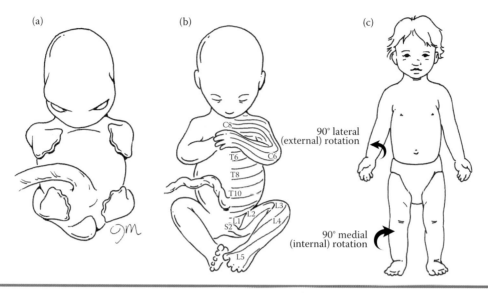

Figure 2.5   Stages of human embryology showing opposite rotations of upper versus lower limbs. (a) At about 19 weeks, the 1st digit of the upper limb (thumb) and lower limb (big toe) form on the cranial side of the developing limb; (b) by about 21 weeks, the upper limb begins to rotate laterally and the lower limb begins to rotate medially; (c) at about 23 weeks, the upper limb has rotated 90° laterally and the thumb is now located on the lateral side (with the arm extended), while the lower limb has rotated 90° medially with the big toe now located on the medial side.

can also be easily seen in photographs showing very early human developmental stages (Plate 2.1). Taking into account the correspondence between the upper and lower limb makes it strikingly easier to learn and compare their anatomy.

Appreciating the evolutionary and developmental history of humans will also help you to understand the full range of movements that can (and cannot) be executed by both limbs. The thigh and the arm can be flexed, extended, adducted, abducted, and rotated, while the forearm and leg can be flexed and extended (moved ventrally and dorsally; the forearm can also be supinated/pronated together with the hand), and their extension is limited by the patella (and ligaments/tendons associated with it, and their surrounding structures) and olecranon, which are both true dorsal structures. Because of the confusion made and incorrect anatomical position described by earlier anatomists, in current human anatomy, the patella continues to be wrongly considered to be an "anterior" structure (i.e., displayed as lying on the ventral side of the body). If one observes the left side of the body of Figure 2.4 displaying the correct anatomical position of humans, then one can easily appreciate that both the patella and olecranon are dorsal structures. In the beginning of Chapter 5, we provide more details about this subject and about which specific limb movements described in most human anatomy textbooks and atlases are consistent or inconsistent with what we know about the evolutionary and developmental histories of the human body.

While we have thus far emphasized the conserved anatomical relationships among most vertebrates, there are some characteristics that are unique to humans or shared with only a few other species. Prominent among us is our bipedal locomotion and the consequent freeing of the forelimbs from body support functions. This trait is shared only with a few other animals, such as birds and some other dinosaurs (phylogenetically speaking, birds are dinosaurs!). In this sense, the anatomical position given in most textbooks is useful because it reflects the reality of our bipedal body although it fails to incorporate our true evolutionary history and ontogeny.

Neither one book nor the collective actions of one class of medical students can change nomenclature that has been used in human anatomy for centuries. Therefore, in this book, we will follow the conventions of anatomical terminology, and thus use the anatomical definitions given in Figure 2.1 (see also more details in the first paragraphs of Chapter 5). In order to be useful for students, the information provided here must be easily comparable to that provided in other textbooks, by their professors, and in their exams. However, contrary to most other anatomical and medical textbooks, we also emphasize the much older biological evolutionary history and **ontogeny (development)** of humans. By doing so, we aim to provide a broader evolutionary, developmental, and pathological context for students to more easily and logically learn and comprehensively understand the human body, while also learning the medical terminology that will allow them to communicate with colleagues, teachers, and patients. This broader perspective will also help students understand embryology because human embryological textbooks tend to use anatomical terms such as "rostral" and "caudal" that are more consistent with evolutionary history (because much of what we know about our development is derived from studies of other tetrapods and vertebrates) rather than the more anthropocentric terms used in human anatomy textbooks/atlases. Therefore, while we use the traditional terminology to describe the structures of the adult human body, we also use the comparative anatomical terminology when we refer to human embryological development.

Chapter 3

Head and Neck

3.1 Development and Organization of the Head and Neck

Some aspects of adult human anatomy are beautifully organized to provide maximal functional efficiency or strength, but it can be difficult for new students to appreciate this organization because adult structures are often highly modified from their original embryonic state. The founder of the discipline of neurology, Ramon y Cajal (1852–1934), was stymied in his attempts to understand the cellular organization of the central nervous system (CNS), and proposed that, "Since the full grown forest turns out to be impenetrable and indefinable, why not revert to the study of the young wood, in the nursery stage...." By looking at embryos, he discovered that the nervous system was composed of individual cells (neurons) and was the first to elucidate the organization of the forebrain and cerebellum.

At first glance, the head seems to lack many of the musculoskeletal features seen in the trunk (Section 2.1), showing: (1) no grossly apparent segmental organization of skeletal structures or peripheral nerves; (2) intersecting—then crossing—passageways for respiration and digestion; and (3) no coelomic cavities separating visceral (e.g., pharynx) from somatic (e.g., body wall) structures. In addition, the head has several **sensory organs** that require specialized innervation and musculoskeletal structures. However, examination of the *embryonic* head reveals that there is, indeed, a set of organizing principles, most of which are similar to those of the trunk. These principles are evident in the spatial relationships among embryonic structures, but become obscured by secondary growth of various parts of the head.

The head has two distinct compartments *with different developmental histories*: (1) the **neurocranium** provides a protective case for the brain and sense organs; in essence, it is a box with two shelves (ear and eye) and one large opening (foramen magnum) for the spinal cord; (2) the **oral cavity**, **nasal cavity**, and **pharyngeal cavity** are surrounded by both rigid and moveable mid-facial and pharyngeal (branchial) arch musculoskeletal structures. The branchial arches are described later in this section.

The facial and pharyngeal bones and cartilages are sometimes referred to as the **viscerocranium** because the pharynx is an endodermal tube, but most of the muscles are striated and voluntary, that is, *somatic* muscles, as opposed to smooth, involuntary *visceral* muscles. However, recent studies have shown that this dichotomy between branchiomeric voluntary muscles versus visceral involuntary muscles is not as clear or well defined as previously thought. For instance, branchiomeric muscles share a common developmental origin with heart muscles (cardiopharyngeal field), and at least part of the esophageal musculature is derived from the branchiomeric musculature (see Box 3.2 and text below).

3.1.1 Migration of Cranial Neural Crest Cells

In Section 6.2, we briefly describe how the early **notochord** plays an important role in inducing the **somites** and overlying neural tissue that gives rise to the vertebrae and ribs. During development, the rostral (superior in adult human anatomy) tissues of the **neural tube** grow so rapidly that they overshoot the tip of the notochord to form the **frontal lobe** of the **forebrain** (Plate 3.1c). The **cranial neural folds**, like the **trunk neural folds**, form a population of cells at their peak, called **neural crest cells** (Plate 3.1a). These cells migrate from their source (Plate 3.1b), like the trunk neural crest cells, but they have many more roles to play than do their analogues in the trunk. In addition to forming **autonomic ganglia** and **melanocytes**, the cranial neural crest cells also form **cartilages** and **bones**, and all the **connective tissues** of the face and jaws, including cells that help pattern the muscles of the **branchial (pharyngeal) arches** (Plate 3.2).

Cranial neural crest cells migrate in discrete streams from the **midbrain** and **hindbrain** regions to form the **facial processes** and **branchial arches** (Plate 3.1b). The single rostral-most stream flows over the rostral tip of the embryo to form the **frontonasal process** that later forms the forehead and middle of the nose. Crest cells move between the primordia of the **optic capsules** (or **optic vesicles**) and the **nasal pits** to form the **lateral nasal processes** (Plate 3.3a and c). The next most caudal streams of cells wrap laterally around the face just caudal to the embryonic optic capsule to form the bilateral **maxillary processes**, which give rise to the **upper jaw** and **hard palate** (Plates 3.1b and 3.3b). The next most caudal paired streams of migrating cells form the **mandibular processes**, which are part of the **1st (mandibular) branchial arch**.

These three streams presage patterns of sensory innervation by the three divisions/nerves of the trigeminal nerve (CN V) complex to the face and jaw. Frontonasal skin is innervated by the **ophthalmic division of the trigeminal nerve (CN V₁)**, the upper jaw, and most of the palate by the **maxillary division of the trigeminal nerve (CN V₂)**, and the lower jaw by the **mandibular division of the trigeminal nerve (CN V₃)**. Most skeletal structures and connective tissues of the face are derived from the migrating cells of the neural crest, whereas the muscle cells associated with these skeletal structures migrate from the anterior-most mesoderm (Plate 3.2).

BOX 3.1   WHY NOT JUST USE MESODERM TO BUILD THE SKULL AND JAWS?

It seems strange to recruit an entirely new neural tube-derived mesenchymal population—**neural crest cells**—to make cartilages and bones that **mesoderm** can generate very well everywhere else in the body. The explanation lies in our ancestry and constraints during their developmental stages. Early chordates—close relatives of vertebrates, in which most of the key features of the body configuration were established—lacked midbrain and forebrain structures and had no **jaws** or segmented **gills**. These features were added (i.e., evolved) later.

So why not simply co-opt head paraxial or lateral mesoderm to do the job? The difficulty here lies in the distinction between shape and content. If you grind up a femur and a mandible, they are biochemically identical. But who would mistake a leg bone for a jaw? The shapes are very different and these differences must be programmed into the progenitors during early developmental stages. As it is generated, paraxial mesoderm is endowed with **positional information**. For example, each **somite** has patterns of gene expression that determine region- and segment-specific anatomical features of axial skeletal structures. The somites that form the **atlas (1st cervical vertebra or C1)** are programmed differently from those, immediately adjacent, that will form the occipital bones. In the trunk, this can be reset slightly; different mammalian species have different numbers of thoracic vertebrae. But in the head region, where formation of the body axis begins, paraxial mesoderm populations cannot reset their spatial programming. Crest cells are derived from the brain and are endowed with the same "positional information" that sets up different regions of the CNS. The genetic changes that expanded the repertoire of cranial neural crest cells and promoted their movements into new, more ventral regions are not fully understood, although some of the **regulatory genes** that endow them with positional information have been identified. Members of the same family of genes provide spatial programming to somites, the CNS, limbs, the gut tube, and much of the urogenital system.

Because cranial neural crest cells contribute to so many structures, genetic defects, or environmental teratogens that affect the proliferation, migration, or interactions of these cells will have widespread effects across the head. For instance, a defect may cause not only hypoplasia (underdevelopment) of oral and facial structures, but also of the ears and the glands that are derived from the branchial arches (thymus, thyroid C cells, and parathyroid glands). Once mesenchymal cells from the neural crest arrive in the arch, these cells interact via intricate back-and-forth signaling with the epidermis to form teeth, nails, hair, salivary, and sweat glands. Thus, ectodermal dysplasia syndromes, which result from genetic disturbances of the signaling process affect all these structures. Moreover, the most posterior stream of migration from the cranial neural crest becomes incorporated into the great outflow vessels of the heart (Plate 3.4). This contribution of neural crest cells to heart development explains why developmental defects of the *cranial* neural crest often result in *heart* defects as well as face, jaw, and ear defects. In a direct clinical example, patients with the congenital muscle disease DiGeorge syndrome must be evaluated for heart defects and immune deficiencies as well as facial defects (see Box 3.2 for further information about developmental links between the heart and head and DiGeorge syndrome).

BOX 3.2   A STRIKING DEVELOPMENTAL CONNECTION BETWEEN HEAD AND HEART MUSCLES

The four-chambered **heart** is initially evident as a single, slightly bent tube underlying the primordia of the 1st and 2nd branchial arches. Mapping studies using transgenic mice revealed that this early heart tube later forms only the left ventricle and atrium. The two "ends" of the tube, the future inflow and outflow tracts, are formed by populations of lateral mesoderm cells that shift from pharyngeal regions deep to the 1st and 2nd branchial arches and elongate the original tube, and neural crest cells that are necessary to separate the common outflow tract into separate aortic and pulmonary channels.

The **first heart field** is a population of early differentiating **cardiac progenitor cells** that arise in the **anterior lateral mesoderm** and give rise to the **early linear heart tube** and, later, to the **myocardium** of the **left ventricle** and parts of the **atria**. The **second heart field** is a population of late differentiating cardiac progenitors that contribute to the developing heart after the linear heart tube stage to give rise to myocardium of the right ventricle and outflow tract and of the inflow tract including parts of the atria. Surprisingly, recent studies have revealed the existence of striking links between the **heart musculature** and the **branchiomeric musculature** (muscles of the branchial arches) (Figure 3.1). Both these musculatures arise from a **cardiopharyngeal developmental field**, which thus includes anterior lateral mesoderm of the first heart field plus contiguous pharyngeal mesoderm that gives rise to second heart field-derived parts of the heart and branchiomeric muscles.

Recent developmental studies reveal a remarkable heterogeneity of the heart and head and blur the interface between head and trunk and between skeletal and cardiac myogenesis. Indeed, adult postcranial structures, including the heart and part of the **neck musculature** (trapezius and sternocleidomastoid), include cells derived from the embryonic cardiopharyngeal field (Figure 3.1, shown in colors). Reciprocally, cephalic structures such as the **tongue muscles** and **infrahyoid muscles** arise from **somitic primordia** located in the trunk, emphasizing the complex developmental and evolutionary history of the head (Figure 3.1, shown in grey). Specifically, recent genetic and developmental studies suggest that: (1) the 1st (mandibular) arch muscles are closely related to the cardiac right ventricle; (2) the left 2nd (hyoid) arch muscles are intimately related to myocardium at the base of the pulmonary trunk; (3) the right 2nd (hyoid) arch muscles are closely related to myocardium at the base of the aorta; (4) the right branchiomeric muscles of the most posterior branchial arches, including the trapezius and sternocleidomastoid, are closely related to the superior vena cava and part of the right atrium; (5) the left

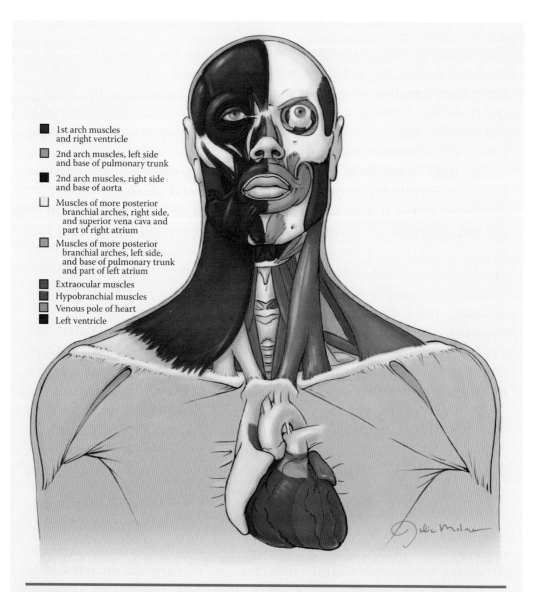

Figure 3.1 **The striking heterogeneity of the human head and neck musculature: It includes at least seven different muscle groups, all arising from the cardiopharyngeal field and being branchiomeric, except the hypobranchial, and perhaps the extraocular, muscles. On the left side of the body (right part of figure), the facial expression muscles were removed to show the masticatory muscles. The seven groups of head and neck muscles are: (1) *1st/mandibular arch muscles*, including cells clonally related to the right ventricle (shown in purple); (2) *left 2nd/hyoid arch muscles*, with cells related to myocardium at the base of the pulmonary trunk (green); (3) *right 2nd/hyoid arch muscles*, related to myocardium at the base of the aorta (red);**

(Continued)

Figure 3.1 (*Continued*)   **(4)** *right muscles of the most posterior branchial arches,* including muscles of the pharynx and larynx and the cucullaris-derived neck muscles trapezius and sternocleidomastoid, with cells related to the superior vena cava and part of the right atrium (yellow); **(5)** *left muscles of the most posterior branchial arches,* with cells related to the base of the pulmonary trunk and part of the left atrium (orange); **(6)** *extraocular muscles* (pink), which are often not considered to be branchiomeric, but that according to classical embryologic studies and recent retrospective clonal analyses in mice contain cells related to those of branchiomeric mandibular muscles; and **(7)** *hypobranchial muscles,* including tongue and infrahyoid muscles that derive from somites and migrate into the head and neck (dark grey). Part of the heart's venous pole is shown in blue, and the left ventricle, derived from the first heart field is shown in brown.

branchiomeric muscles of the most posterior branchial arches are closely related to the base of the pulmonary trunk and part of the left atrium.

Recognition of the close relationship between specific head and heart muscles sets the stage for future biomedical studies relevant for human medicine. One of the most important questions in the **pathology** of **muscle diseases** is why many myopathies preferentially affect a specific subset of muscles and whether these etiologies are linked to disparate embryonic histories. **DiGeorge syndrome** is one of the most frequent **human congenital syndromes**. Its clinical features include **cardiovascular defects** and **craniofacial anomalies**, highlighting the frequent linkage of cardiac and craniofacial birth defects due to their anatomical proximity during early embryogenesis and overlapping progenitor population. Future meta-analyses of human syndromes may reveal pathological relationships between the specific branchiomeric muscles and cardiac regions shown in Figure 3.1 and thus help to understand regional congenital heart defects. Applications of this research may include mapping human disease polymorphisms to relevant enhancers in the cardiopharyngeal network and using branchiomeric myoblasts for heart muscle repair.

3.1.2 The Branchial Arches

The branchial arches, often designated "**pharyngeal arches**," are: the **1st arch** or **mandibular arch**; the **2nd arch** or **hyoid arch**; and two more caudal arches. Our ancestors had many more branchial arches, each of which formed gills, but these have been lost or highly transformed during vertebrate evolution (Plate 3.2). Knowledge of **branchial arch development** makes it easy for students to understand the **innervation patterns of the head**.

Within each **branchial arch** neural crest cells form a central core of cartilage (one or a series of several) and surrounding connective tissues and mesodermal cells coalesce to form an artery. Each arch is penetrated by mesodermal cells that form **skeletal muscles** that will be innervated by the cranial motor and sensory nerves specific to that arch. This relationship, largely unchanged during vertebrate evolution is retained regardless of any subsequent displacements of the muscle. The muscles associated the branchial arches are called **branchiomeric muscles** and will be described in detail in Section 3.4.

The **1st (mandibular) branchial arch** is the embryonic structure that gives rise to the **upper jaw** and **lower jaw** including bones, muscles and connective tissues, and upper and lower dentition (teeth) (Plate 3.2). Early in embryonic development, this arch subdivides into two major, paired (left and right) tissue swellings, the ventral mandibular prominence and the dorsal maxillary prominence (Plates 3.2 and 3.3). The **mandibular prominence** gives rise to all lower jaw bones, muscles, and associated connective tissues. Its contributions vary somewhat among mammals and vertebrates in general, but in humans the **maxillary prominence** contributes to the upper jaw, cheeks, and secondary palate, and associated connective tissues. These paired prominences fuse with crest-derived populations of the **frontonasal prominence** to complete formation of the midface. Failure of any or all of these prominences to properly contact and fuse will result in a congenital clefting defect of the palate, lip, or face. These clefts, which can be caused by genetic and/or environmental factors and occur as often as 1 per 800 births, can be surgically corrected. The earlier after birth surgery is performed, the better the prognosis.

At the junction between the maxillary and lateral nasal prominences, a groove forms, invaginates, and separates from overlying surface ectoderm, forming an internal tube (**nasolacrimal duct**) that runs from the orbit to the nasal cavity. As the left and right maxillary prominences continue to elongate, they eventually contact and then fuse with the medial nasal prominences, enveloping the nasal pits and creating the **external nares**.

The exact sites and extent of fusion between maxillary and medial nasal prominence vary across mammals, and even within a species, as demonstrated by the different "nose prints" seen in domestic species such as cats, dogs, and horses. The fact that each individual human face looks unique, despite being formed via the same embryonic processes, is probably due to subtle differences in rate and direction of programmed tissue outgrowth (e.g., cell proliferation) that "tweak" the shape of a nose or a chin. These processes are directed in large part by quantitative variations in patterns of growth-promoting genes that we inherit from our parents (e.g., "mom's nose, dad's chin").

BOX 3.3   FACIAL DIVERSITY

What mechanisms produce the many different faces found in the animal kingdom in general and of humans in particular? This question is being dissected by molecular analyses of gene expression patterns within the **frontonasal prominence** and **maxillary prominence** before and following their early expansions and fusions. For example, analyses of embryos from several species of Galapagos finches, which have evolved different beak shapes in association with different feeding behaviors, reveal that different levels of expression of just a few growth-promoting/retarding genes in beak precursors (neural crest cells, surface epithelium) and the underlying prosencephalon can account for most of these differences. Mis-expressing these genes in chick embryos reproduces some of these naturally occurring variations. While these findings nicely explain variation in facial morphology, the question of how behavioral adaptations and changes in intestinal structure and function coevolved with new beak shapes has not been answered.

The **mandibular cartilage** is often called **Meckel's cartilage**, because it was first described by the German anatomist Johann Friedrich Meckel. Meckel's cartilage extends from near the midline of the chin all the way back to the developing middle ear (Figure 3.2), and is a source of genetic instructions to guide ossification by surrounding neural crest cells. In human embryology, the more rostral bar of Meckel's cartilage is called the **palatopterygoquadrate bar**. The perichondrial membrane surrounding the middle portion of Meckel's cartilage becomes the **sphenomandibular ligament**, which connects the **lingula of the mandible** to the **spine of the sphenoid** (Plate 3.28a). In the early evolution of mammals, the lower jaw changed from having two bones (quadrate, articular) between the biting element (dentary) of the lower jaw and the skull (squamosal bone) to having the dentary (the mammalian mandible) articulate directly with the squamosal. Freed of their roles in jaw movement, the quadrate and articular, both derived from the proximal (caudal) end of Meckel's cartilage, were co-opted to serve as parts of the vibration-transmitting apparatus of the middle ear, the **malleus** ("hammer") and **incus** ("anvil"). All of these elements, including the **squamosal**, are part of the 1st branchial arch. Prior to these changes, sound was transmitted from the tympanic membrane to the cochlea by a single bone, the columella (mammalian **stapes**), which is

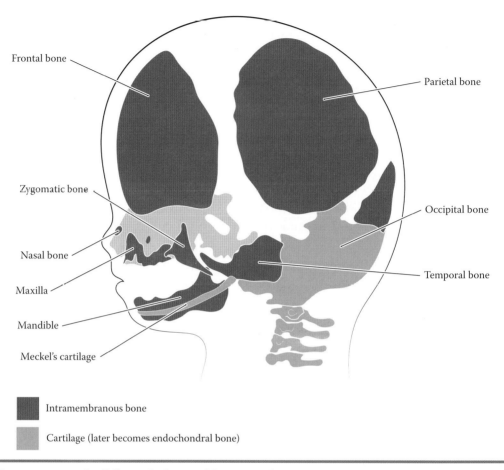

Frontal bone

Parietal bone

Zygomatic bone

Occipital bone

Nasal bone

Maxilla

Temporal bone

Mandible

Meckel's cartilage

Intramembranous bone

Cartilage (later becomes endochondral bone)

Figure 3.2 **Early differentiation and intramembranous ossification of human skull bones.**

derived from the 2nd branchial arch (Plate 3.45a and c). Despite these evolutionary and developmental changes in human adults, all the muscles that originally form in the mesenchyme of the mandibular and maxillary processes are still innervated by the nerve of this arch: the **trigeminal nerve (CN V)** (Plate 3.15). That is why, in addition to the **masticatory muscles** and the **anterior digastric**, muscles such as the **tensor veli palatini** (attached to the palatine aponeurosis) and the **tensor tympani** (attached to the malleus) are also innervated by the trigeminal nerve (Plates 3.36 and 3.45).

The sphenoid bone is a large, "winged" or butterfly-shaped bone in the central region of the skull that is only visible externally at pterion (a skull landmark defined in Section 3.2.1.1). It is marked on its dorsal surface by the fossa in which the pituitary gland sits. This bone resides at the boundary

BOX 3.4   DEVELOPMENT OF CARTILAGE AND BONE

The entire axial and, except for the clavicle, appendicular skeleton develops initially as **cartilage**. In the head, the walls, and floor of the braincase are also initially cartilaginous. **Chondrogenesis** (cartilage formation; chondro = cartilage) always begins with the aggregation of mesenchymal cells in response to signals from nearby tissues. Aggregated **chondroblasts** initiate the synthesis of cartilage matrix-specific molecules such as collagen type II and chondroitin sulfate proteoglycan. Shortly after the onset of differentiation, cartilage becomes enveloped with a fibrous surrounding sheath called the **perichondrium**. This event marks the end of incorporation of new cells into the aggregate, and all further growth occurs by a combination of chondroblast proliferation within the perichondrium and cartilage, and of enlargement (hypertrophy) due to increases in the size of cells and volume of matrix. One hallmark of cartilage is that it is poorly vascularized, and most nutrients arrive via diffusion from surrounding blood vessels.

In contrast, **bone** arises by two distinct processes. In skeletal structures formed initially as cartilage, ossification begins within or along the margins of the cartilage. This process is called **endochondral ossification**. **Osteoblasts** (bone-forming cells; osteo = bone) produce a type I collagen-rich matrix. A second bone-forming strategy is to initiate ossification without a preexisting cartilage; this is **intramembranous ossification** (also referred to as dermal bone formation). The term "intramembranous" is not strictly accurate, as the initial stages of intramembranous ossification involve aggregations of mesenchymal cells, which may or may not be associated with a membranous layer. Intramembranous ossification is the strategy of choice in the jaws and the roofing bones of the skull.

Why do most skeletal structures require this two-step process? The answers are both developmental and functional. The genetic blueprints for *when* skeletal elements will form, for *what their initial shapes* will be, and *where joints will form* are executed in mesenchymal populations that form cartilage. This ancient ritual has not been transferred to bone. Places where intramembranous bones develop are highly specialized environments in which "decisions" about when, what, and where tissues form are largely imposed on mesenchymal structures by surrounding tissues. Functionally, bone is a continuously remodeling tissue that is especially sensitive to mechanical forces placed on it. Within bone, trabeculae become aligned to maximally resist whatever forces they experience, for example, *compression* on vertebrae due to the weight of the body while standing, or *tension*, for example, pulling forces, at the sites of muscle attachment when muscles contract. Additionally, bones become thicker in response to higher mechanical forces: For example, an X-ray will show that the bones in the serving arm of tennis players are thicker than the bones of the other arm and astronauts who spend time in zero gravity must exercise rigorously to avoid extreme bone loss. Cartilage does not have the capacity to adapt to mechanical forces, but has greater flexibility and compressibility, making it an ideal surface for covering bones at joints.

The roof of the **skull** grows rapidly during fetal stages and, in humans, for several years after birth. This rapid growth is necessary to accommodate the extensive increases in brain size as neurons mature and become myelinated. The **skull roof** is dominated by the large, flat **parietal** and **frontal** bones, which differentiate directly as bone (= intramembranous ossification). The sites at which ossification is initiated are determined by signals emanating from the **brain**. These osteogenic sites expand between the **dura** (outermost of the three **meninges**, the membranous layers that cover the brain, and spinal cord) and the overlying skin, forming large sheets of **osteocytes**. The sites where these plate-like sheets of bone make contact (and often overlap) are **sutures**, which are described at the beginning of Section 3.2.

When most people think of bones they picture the typical museum specimens, which are only the extracellular inorganic matrix component. However, living bone is as dynamic a tissue as any other, although it operates under the constraints of having to accommodate both compression and tension. As with so many other systems, the developing fetus must prepare for both (even though there are few compressive forces in utero, except those produced by muscle contraction), especially in those species that begin walking within moments of birth.

between neural crest- and mesoderm-derived skeletal components of the skull, with its rostral part and greater wings being of crest origin and the remaining, caudal portion from mesoderm. The anterior-most portion of the palatopterygoquadrate bar is incorporated into the **greater wing of the sphenoid**, the larger and more cranial of the "butterfly wings" (Plates 3.10

and 3.12). The **pterygoid plates** (Plates 3.8a and c and 3.10), then form by intramembranous ossification. As noted in the previous paragraph, the posterior portion of the palatopterygoquadrate bar ossifies endochondrally as the **incus**, the anvil-shaped middle ear bone (Plate 3.45a and c). In early development, the **malleus-incus** (or "hammer-anvil") joint thus functions as the primary jaw joint of the 1st branchial arch. During development, the mandible acquires a condylar process that contacts the growing **zygomatic process of the temporal bone**, forming the secondary, definitive adult jaw joint, which is formally named the **temporomandibular joint** to acknowledge the connection of the temporal bone with the mandible (Plate 3.28).

BOX 3.5   DEVELOPMENT OF JOINTS: USE THEM OR LOSE THEM

Any contact between separate skeletal elements constitutes a **joint**. The articulations between long bones are moveable joints in which the apposing bones are coated with a layer of cartilage, and the space between them is fully enclosed by a dense, fibrous **joint capsule**. These highly mobile joints are called **synovial joints** and the fluid contained within the joint is **synovial fluid**. During limb development, cartilages form proximally to distally, and cartilaginous rods typically span the sites where joints will form. The rods branch at sites where more than two bones are joined, for example, the future elbow (knee) and wrist (ankle) locations or simply elongate and secondarily become separated, for example, within the digits.

Wherever a joint forms within an existing cartilage, a band of immature **chondrocytes** undergoes localized programmed cell death. The dying chondrocytes are replaced by **fibroblasts** and a new **perichondrium** is established around the ends of the cartilages. Cells of the former perichondrium at these degenerating zones persist and differentiate as the **fibrous connective tissues** of the joint capsule. These connective tissues provide mechanical stability and also form part of the seal that retains synovial fluid within the joint. Later, **articular cartilage** will form on the surfaces of the bones facing the joint.

Interestingly, if joints are immobilized during fetal stages, remodeling of the cartilaginous lining of the joints and formation of connective tissues within the joint are disrupted. This disruption often leads to functional fusion of the joint, a condition called **ankylosis**. The resulting abnormal limb posture is **arthrogryposis**. The mechanisms by which morphogenesis of joints is disrupted in the absence of biomechanical forces are not well understood. Recent studies have shown that patterns of gene expression in both the emerging epiphyseal cartilage and the connective tissues in and around the joint are affected by changes in biomechanical activity. Genes affected include those whose products are **mitogenic growth factors**, such as PTHrP (parathyroid hormone-related protein), which is also necessary for tooth eruption, FGF2 (fibroblast growth factor 2), and several **collagens**.

The **2nd branchial arch** also has several skeletal derivatives. Like the 1st arch-derived Meckel's cartilage, its embryonic cartilage (**Reichert's cartilage**, which, as you might have guessed, was named after a German anatomist named Reichert, Karl Bogislaus) also has two parts early in development. The posterior-most portion gives rise to the **stapes** (stirrup-shaped) ossicle of the middle ear and to the **styloid process of the temporal bone** (apparently so named due to its association with time, as grey hairs first appear at the temple) (Plates 3.8 and 3.45). The inferior-most portion becomes the **lesser horn** and **superior rim of the hyoid body**. The **stylohyoid ligament** (Plate 3.13) is formed between the two portions of this embryonic cartilage. The muscles of the 2nd branchial arch are innervated by the **facial nerve (CN VII)** and include the **stapedius muscle**, the **stylohyoid muscle**, and the **posterior digastric muscle** (Plates 3.26, 3.27, 3.45, and 3.16). These muscles all attach to derivatives of Reichert's cartilage. The **muscles of facial expression**, often simply called the **facial muscles**, are an exception to this conserved muscle-connective tissue relationship (Plate 3.27). Precursors of these muscles spread beneath the skin of the head and neck, but owing to their 2nd branchial arch origins, remain innervated by the facial nerve (CN VII). Included in this category are the superficial muscles of the scalp—**occipitalis** and **frontalis** (often designated as a single muscle **occipito-frontalis**) and **auricular muscles** (often called the **extrinsic ear muscles**)—muscles of the face proper (including the **buccinator**) and of the neck (**platysma**).

Endochondral ossification of the **3rd branchial arch** cartilage produces the **greater horn of the hyoid bone** and the **inferior rim of the hyoid body** (Plate 3.13). The main nerve of this arch, the **glossopharyngeal nerve** (CN IX) innervates only one muscle, the **stylopharyngeus** (Plate 3.17). This muscle is attached to the **styloid process**, dorsally, then spreads out along the internal wall of the middle pharyngeal constrictor (which attaches to the greater horn of the hyoid bone, a 3rd arch derivative).

The 4th branchial arch forms the thyroid and epiglottis cartilages of the larynx and, according to various studies, also the lesser horn of the hyoid bone. Muscles arising from the 4th branchial arch (and perhaps adjacent mesoderm) are mainly innervated by branches of the **accessory nerve (CN XI)** and **vagus nerve (CN X)**. CN XI innervates the **trapezius** and **sternocleidomastoid** muscles, whereas somatic branches from CN X innervate intrinsic laryngeal and pharyngeal muscles (Plate 3.2). Without going into detailed embryology, consider that the muscles innervated by the **vagus nerve** (CN X) (Plate 3.18) can be divided into two groups. The more rostral muscles can be considered as a nearly continuous sheath of muscle surrounding the pharynx. The **superior, middle, and inferior constrictors** all follow a pattern of arising from a midline raphe and sweeping forward to attach to structures lateral to the **pharynx** (the **pterygomandibular raphe**, the **greater horn of the hyoid**, and the **thyroid cartilage** and **cricoid cartilage**) (Plate 3.35). If the inferior attachment of the inferior constrictor is followed past its attachment on the oblique line of the thyroid cartilage, it is easy to envision the **cricothyroid muscle** as an extension of this sheath (Plate 3.35). That is why this latter muscle is also innervated by a branch of the vagus nerve (the external laryngeal nerve) that also innervates the inferior constrictor. The **levator veli palatini** can likewise be envisioned as an extension of the same sheath at the rostral end (Plate 3.36b). The **superior constrictor** muscle attaches at the **pharyngeal tubercle of the occipital bone** and at the **pterygoid hamulus** and pterygomandibular raphe, and is also connected to the tongue and to the mandible. The attachment of this muscle to the skull is completed by the **pharyngobasilar fascia**, forming a C-shaped attachment to the skull (Plate 3.36b). Running through the pharyngobasilar fascia is the **pharyngo-tympanic tube**, taking a slip of the rostral-most portion of the muscle sheath with it. If you consider that the **palatal shelves** were originally on the sides of the developing **oral cavity**, before their elevation and merging at the midline, then the continuity of the insertion makes sense. In this regard, it makes sense that the vagus nerve innervates the **levator veli palatini** (which originates on the temporal bone just posterior to the opening of the bony pharyngo-tympanic tube and inserts on the superior surface of the midline of the soft palate) (Plate 3.2).

Most textbooks refer to six branchial arches in mammals, but studies performed in the last decades by Noden and colleagues indicate that, during the evolution of amniotes (reptiles and mammals), the caudal two arches were lost, leaving four—not six—true branchial arches. In this view, the "6th" aortic arch is not seen as a true aortic arch (see Plate 3.5), but rather as a new anastomosis between the multiple branches from the paired dorsal aortae and the pulmonary arteries, which are initially branches from the ventral part of the 4th aortic arch. Formation of direct connections between the pulmonary arteries and the dorsal aortae is essential to the embryo, as it allows oxygenated blood from the placenta to reenter systemic circulation without having to pass through the nonfunctional lungs. This connection is retained only on the left side as the **ductus arteriosus**. Whereas in rare cases, a small artery in the position where a 5th aortic arch might occur is found, according to Noden and colleagues it is more likely an anomalous remodeling of the aortico-pulmonary plexus.

The **hypoglossal nerve (CN XII)** arises from the caudal part of the hindbrain, which lies in the caudal part of the neurocranium (occipital) derived from somites. This nerve innervates the **hypobranchial muscles (tongue muscles** and **infrahyoid muscles)**, which are not branchiomeric muscles derived from **pharyngeal mesoderm** but instead develop from myoblasts derived from **somites** (Box 3.2; Figure 3.1; Section 3.3).

BOX 3.6   REGISTRATION: JUST A COINCIDENCE?

Is the fact that muscles (from mesoderm), skeletal structures (from neural crest), and motor innervation (from hindbrain) for each branchial arch all arise in the same axial neighborhood (Plate 3.2) just a coincidence? To test this, transplants were done in which each of these progenitor populations were moved from one axial location to another. Neural crest cells that normally would form the 1st branchial arch (= lower jaw) were grafted in place of crest cells that normally would have occupied the 2nd branchial arch (= hyoid cartilages). The grafted crest cells dispersed normally for their new location, but in the position of the 2nd arch formed a lower jaw. They knew their spatial identity before leaving the neural folds! Moreover, paraxial mesoderm cells that moved into the rebuilt 2nd arch formed jaw, not hyoid, muscles.

How is this programming achieved? Mapping of gene expression patterns in neural crest cells reveals that 2nd arch crest cells express several *Hox* genes, but 1st arch crest cells do not. In mice, in which some of the 2nd arch *Hox* genes were nullified, the crest cells in this arch form mandibular rather than hyoid structures.

Thus, registration among the several branchial arch progenitors is not a coincidence, but rather reflects communications directing spatial organization among participants, with the connective tissue-forming population at the top of the information dissemination network. Full expression of both spatial programming and cell lineage determination requires signals from other nearby tissues, including the pharyngeal endoderm and surface epithelium, but these signals act on populations whose programming was initiated before entering the branchial arch environment.

3.2 Head and Neck Skeletal System

3.2.1 Cranium

Topologically speaking, the adult human **skull** includes the **cranium** and **mandible**. In this section, we describe the skull by region, beginning with the superior aspect (calvaria) and moving sequentially to the inferior, anterior, lateral, and interior aspects.

3.2.1.1 Calvaria

The skull cap, formed by parts of the frontal, parietal, and occipital bones, is called the **calvaria**. The cranial bones are joined by sutures, which are immobile joints of interlocking bone. The **parietal bone** includes the posterior portions of the **superior and inferior temporal lines**. The suture between the parietals and the **occipital bone**—which contains the **external occipital protuberance**—is called the **lambdoid suture** (not because of anything to do with young sheep, but because the suture's path resembles the divergent shape of the Greek letter, *lambda*). The point where this suture meets the **sagittal suture** (named of course for its position in the sagittal plane; see Figure 2.1) separating the left and right parietal bones, is called **lambda** (Plate 3.7b). The suture between the parietals and the frontal bone is called the **coronal suture** (because of the suture's position in the coronal plane). The area where the sagittal and coronal sutures meet is called **bregma**, an odd-sounding moniker that apparently denoted the top of the head to the ancient Greeks who named this skull landmark. The intersection of the frontal, sphenoid, temporal, and parietal bones is called the **pterion**, which exhibits at least four variations in terms of which of

BOX 3.7   CRANIOSYNOSTOSIS

Craniosynostosis (Figure 3.3) is the condition resulting from premature closure of any cranial suture. This defect occurs when osteogenic stem cells stop dividing and differentiate as bone, thereby arresting the growth capability of a suture. Continued expansion of the brain pushes against the skull roof and sides, *causing excess growth at the other sutures.* This unequal growth results in marked asymmetries in the skull and, often, compression-induced **necrosis** of parts of the brain. For example, in **sagittal synostosis**, the head cannot expand in width and therefore becomes

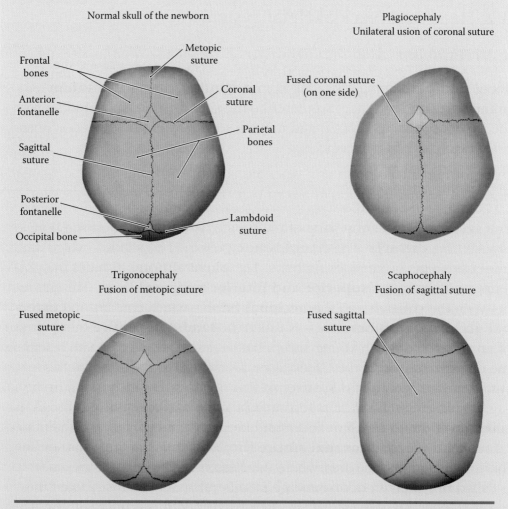

Figure 3.3   Different forms of craniosynostosis, as depicted in sketches of the top of the head.

atypically elongated. Some premature fusion of sutures is evident in 1:2500 human infants with males affected twice as frequently as females.

Clinically, it can be a challenge to distinguish mild craniosynostosis from transient, externally-imposed deformations of the skull, for example, during passage of the neonate through the birth canal or occasionally as a result of an infant sleeping on its side. Craniosynostoses are categorized according to the suture affected and also the degree of premature fusion. **Plagiocephaly**, in which the coronal or lambdoid suture closes prematurely on only one side, is the most common. In children with plagiocephaly, the forehead and brow on the affected side are flattened, and the contralateral structures are excessively prominent. Craniosynostosis is present in over 100 syndromes, many of which include abnormalities in the formation of the skull base and growth of long bones. Fewer than 10% of patients with craniosynostosis have a familial history of the disease. A percentage of **sagittal synostoses** are the result of compression forces exerted by the uterus during the last month of gestation; typically, these pressures create only focal zones of suture closure, and do not require surgical intervention.

these four bones contact each other. The suture between the temporal and parietal bones is called the **squamosal** (it's a stretch to imagine but this name derives from "plate" or "scale"-like, because the suture separates the parietal bone from the squamosal region of the temporal bone; this region contrasts with the petrous, or "stone-like," region of the temporal bone) suture (Plate 3.8a).

3.2.1.2 Basicranium

As will be discussed in detail in Chapter 6, all **somites** look similar, yet the vertebrae that they form have unique shapes in each part of the body. The first four true somites do not form vertebrae at all; they form the **occipital bone**. These somites initially look like all the others, but their **sclerotomes** (a division of the somite that also forms vertebrae) all fuse together to form a single elongated body, the **basioccipital**, around the notochord. Lateral and dorsal parts of the occipital cartilage are also formed by fusion of adjacent sclerotomes and all the parts of the occipital cartilage undergo endochondral ossification to form the **occipital bone**. Instead of articular facets, the caudal surface of the occipital bone has large, convex **occipital**

condyles. The articulation between the occipital condyles and the 1st cervical vertebra, called the atlas (described in the next paragraph), restricts movement to extension/flexion (i.e., this is the "yes" joint) (see also Section 6.1 for descriptions of form, function, and development of vertebrae). The adult occipital bone forms much of the back and base of the skull. In addition to the **occipital condyle**, it has a protuberance called the **pharyngeal tubercle** to which the pharyngeal raphe attaches. Openings in the occipital bone include the **foramen magnum** (the "big hole" through which pass the **medulla oblongata**, **spinal roots of accessory nerve (CN XI)**, and **vertebral arteries**), and the paired **hypoglossal canals** for the passage of the **hypoglossal nerve (CN XII)** (Plate 3.11).

Internally, the occipital bone is marked by grooves for several **dural sinuses**: the **sigmoid sinus** ("S-shaped"), **superior sagittal sinus**, and the **transverse sinuses**. It also contains fossae for parts of the brain: the **fossae for the cerebellum** and the **fossae for the occipital poles of the cerebral hemisphere** (Plates 3.10 through 3.12). The **atlanto-occipital joint** is made by the occipital condyles and the superior articular facets of the atlas bone, which is part of the vertebral column (the **atlas** bone is named after Atlas, the mythical man who balanced the world on his shoulders, as the atlas balances the head on the neck). The atlas includes the **anterior arch**, **superior articular facet**, **transverse process**, and **posterior arch** (Plate 3.31b and c). Note that the **transverse ligament of the atlas** holds the dens to the anterior arch of the atlas. The **retropharyngeal space** is situated posterior to the cervical viscera and extends from the **basilar part of the occipital bone** to the **superior mediastinum**.

The **foramen magnum** is clinically important because any defect in the formation or enlargement of this foramen or of the spinal canal through cervical vertebrae 1 and 2 will compress the spinal cord, producing clinical signs of ataxia (poor coordination) and weakness. The foramen magnum is also evolutionarily important because it is typically located at the back (posterior region) of the skull in quadrupedal mammals and other vertebrates, but uniquely positioned at the base (inferior region) of the skull in bipedal animals, notably humans (including babies who are still crawling—even they can sit upright with their torsos beneath, not behind, their heads).

3.2.1.3 Anterior Cranium

Shifting from the base and back of the skull to its front, the **frontal bone** includes the following landmarks and structures: **glabella**, the **superciliary** arch, **supraorbital notch** (**foramen**), **foramen cecum**, **orbital surface**, **lacrimal fossa**, and **supraorbital margin** (Plates 3.7a and c, 3.8a and c, and 3.12). The junction between the frontal bone and the **nasal bones** is called the **nasion**. The roof of the skull grows rapidly during fetal stages and, in humans, for several years after birth. This rapid growth is necessary to accommodate the extensive increases in brain size as neurons mature and become myelinated. The **frontal suture** or **metopic suture** (meaning "among") is the suture between the two frontal bones and is usually closed in the adult but palpably open in a newborn human baby, which helps the infant's head to safely compress to ease its movement through the birth canal.

The **orbital margin** is formed by the frontal bone superiorly, **zygomatic bone** laterally (which includes the **frontal process** and **temporal process**) and the **maxillary bone** (or maxilla) medially, which also spans a good portion of the skull. The maxilla includes a **frontal process** along this orbital margin, an **alveolar process** above the teeth, an **anterior nasal spine** located at the bottom of the nasal region protruding out, an **incisive foramen**, and **palatine process** along the roof of the oral cavity, an **anterior lacrimal crest** bordering the anterior portion of the lacrimal groove, an **infraorbital groove** located on the floor of the orbit, and an **infraorbital foramen** just inferior to the orbital margin. The **anterior nasal aperture** is bounded by the nasal bones and maxillae. The **nasal septum** and the **lacrimal bone** (which includes the **posterior lacrimal crest** and **lacrimal groove**, all so-named because of their proximity to the lacrimal duct, which drains the tears of happiness that you will shed on completing your degree) can be seen in a frontal view of the anterior nasal aperture (Plates 3.7a and c and 3.8a and c).

BOX 3.8 EASY WAYS TO REMEMBER: NAMES OF BONE PROCESSES

Students tend to confuse the names of the processes of bones. However, the people that established these names generally followed a very logical nomenclature. So, in most cases, the name "process *X* of the *bone A*" simply refers to a prominence of bone A that meets bone X. For instance, the "*zygomatic* process of the *temporal bone*" refers to a prominence of the *temporal bone* that runs toward, and meets, the *zygomatic bone*. Accordingly, the "*temporal* process of the *zygomatic bone*" refers to a prominence of the *zygomatic bone* that that runs toward, and meets, the *temporal bone*.

3.2.1.4 Lateral Cranium

The external surface of the **temporal bone** includes the **zygomatic process**, **mastoid process** (mastoid means "breast-like" in Latin, in reference to the rounded shape of this process), **styloid process** ("stick-like" or "pointed"), **mandibular fossa**, **external acoustic meatus**, **articular tubercle**, **carotid canal** (for internal carotid artery), the **bony portion of the pharyngotympanic tube**, the **jugular fossa**, and the **stylomastoid foramen** (Plates 3.8a and c and 3.10). The **temporal fossa** is a depression formed by the parietal, **frontal**, **squamous part of the temporal** bone, and **greater wing of the sphenoid**. It serves to expand the area for attachment of jaw closing muscles. The **zygomatic arch** (colloquially known as the "cheekbone") is formed by the **zygomatic process of the temporal bone** and the **temporal process of the zygomatic bone**. The **infratemporal fossa** lies inferior to the zygomatic arch and includes or adjoins the following structures: The **pterygomaxillary fissure** lies between the **lateral plate of the pterygoid process** of the **sphenoid bone** and the **infratemporal surface of the maxilla**; the **pterygopalatine fossa** lies at the superior end of the pterygomaxillary fissure, and the **sphenopalatine foramen** is an opening into the **nasal cavity** (Plates 3.8 and 3.9);

**BOX 3.9   EASY WAYS TO REMEMBER:
PTERYGOMAXILLARY FISSURE, PTERYGOPALATINE
FOSSA, AND SPHENOPALATINE FORAMEN**

Students tend to confuse the names **pterygomaxillary fissure**, **pterygopalatine fossa**, and **sphenopalatine foramen**. An easy way to remember them is to begin with the most inclusive structure (the pterygomaxillary fissure) and move to the lease inclusive structure, keeping part of the original word with each step. So, between the maxilla and pterygoid region logically lies the *pterygomaxillary fissure*, which is the most inclusive structure of the three; then this fissure leads to the *pterygopalatine fossa* (so, only change the "maxillary" part at the end of the word to "palatine"), and then the less inclusive structure is named the *sphenopalatine foramen* (change the "pterygo" at the beginning to "spheno"). It is also useful to recall that a fissure is as described (crevice-like; might be an opening, or not), whereas a fossa is a depression and a foramen ("window") is a hole.

the **inferior orbital fissure** lies between the **maxilla** and the **greater wing of the sphenoid bone**, which contains the **foramen ovale** and **foramen spinosum**.

3.2.1.5 Cranial Cavities

The centrally-located **ethmoid bone** is a light, spongy bone (ethmos = "sieve," and thus ethmoid means "sieve-like") that acts rather like a keystone anchoring the cranial bone with the facial bones. The intricate and vaguely rectangular ethmoid bone includes the **crista galli** and **cribriform plate**—with foramina for bundles of the **olfactory nerve (CN I)** (Plate 3.2). From the same superior view of the cranial base that shows the ethmoid, one can see that the **sphenoid bone** includes the **lesser wing, anterior clinoid process, greater wing, carotid groove** (for the internal carotid artery), and the **sphenoid body**, which comprises the **jugum, chiasmatic (optic) groove**, and the **sella turcica (tuberculum sellae, hypophyseal fossa, dorsum sellae, and posterior clinoid process)** (Plates 3.11 and 3.12). The sphenoid bone also encloses the **optic canal**—for **optic nerve (CN II)** and ophthalmic artery; the **superior orbital fissure**—for **oculomotor nerve (CN III), trochlear nerve (CN III), ophthalmic nerve (CN V$_1$), abducens nerve (CN VI)**, and superior ophthalmic vein; the **foramen rotundum**—for **maxillary nerve (CN V$_2$)**; **foramen ovale**—for **mandibular nerve (CN V$_3$)**; and the **foramen spinosum**—for middle meningeal artery and vein (Plate 3.2). Note that the **foramen lacerum** lies between the sphenoid and temporal bones, the **anterior cranial fossa** is separated from the **middle cranial fossa** by the sphenoidal crests and limbus, and the **posterior cranial fossa** is also separated from the middle cranial fossa by the **superior border of the petrous part of the temporal bone** and by the dorsum sellae. On the medial wall of the orbit between the sphenoid and the frontal bones lie the **anterior ethmoidal foramen** and the **posterior ethmoidal foramen** through which pass the anterior and posterior ethmoidal arteries, veins, and nerves, respectively. The part of the ethmoid bone that forms the medial aspect of the orbit is so thin that is called the **lamina papyracea** ("papyrus," the thin plant-derived paper; "lamina" means layer).

The **nasal cavity** includes the bones of the **lateral nasal wall** and the **nasal septum**. The lateral nasal wall includes the **cribriform plate** superiorly, the intricately swirled **superior nasal concha** (concha = "shell shaped," to increase surface area for air circulation, warming, and decontamination;

inferior to the **sphenoethmoidal recess**) and the **middle nasal concha** (inferior to the **superior meatus**) of the **ethmoid bone**. The **lacrimal bone** (a small fragile bone descending from the orbit) and the **inferior nasal concha** (inferior to the **middle meatus** and superior to the **inferior meatus**) are the other two bones populating the lateral nasal wall. Posteriorly, one can find the opening of the **sphenoidal sinus**, **sphenoid body**, **medial plate of the pterygoid process** of the **sphenoid bone**, the **sphenopalatine foramen** as described earlier, and the **perpendicular plate** and **horizontal plate** of the **palatine bone** (Plates 3.9 and 3.38). The **vestibule** (meaning, "the space between the entrance and the interior") is anterior to the inferior meatus and the **atrium** ("open space") is anterior to the inferior meatus. The position of **nasal choanae** (the caudal openings of the nasal passages) shifts caudally with the formation of the secondary palate.

During development, additional outgrowths of epithelium lining the nasal cavity will form sinuses within the maxilla, ethmoid, frontal, and sphenoid bones. Generally, these sinuses do not expand until after birth and are proportionately larger in mammals with heavier skulls (which makes sense because these air-filled cavities serve to lighten the bone). Lateral to the inferior concha lies the opening of the **nasolacrimal duct**, lateral to the middle concha lies the **semilunar hiatus** (with openings of the **frontal sinus, anterior ethmoidal cells**, and **maxillary sinus**) and the **ethmoidal bulla** (with the opening for **the middle ethmoidal cells**). The **opening of the posterior ethmoidal cells** lies in the superior meatus and the **opening of the sphenoidal sinus** lies in the sphenoethmoidal recess (Plates 3.9 and 3.38). Inferior to the **nasal bone** lies the **lateral nasal cartilage**, which is an extension of the **septal cartilage**. The septal cartilage separates the right and left nasal cavities and forms the anterior aspect of the **nasal septum**, lying medially to the **alar cartilage** that gives shape to the nostril (Plate 3.9b). Apart from the septal (or "dividing") cartilage, the nasal septum is formed by the **perpendicular plate of the ethmoid bone** and the **vomer** (Plate 3.9a and b). Therefore each nasal cavity is bordered superiorly by part of the nasal septum and parts of the nasal bone, cribriform plate of ethmoid bone, and sphenoid bone; inferiorly by the palatine process of the maxilla and horizontal plate of the palatine bone; medially by the nasal septum; and laterally by the lacrimal bone, maxilla, ethmoid bone, perpendicular plate of the palatine bone, and inferior nasal concha.

The **hard palate** is a conglomerate of many structures making up the roof of the oral cavity and these structures also seamlessly contribute to nearby regions. These include the **incisive foramen** located just posterior to the incisors teeth, **alveolar processes** superior and adjacent to the teeth and **palatine process** of the **maxilla**, the **horizontal plate, perpendicular plate, greater palatine foramen**, and **lesser palatine foramen** of the **palatine bone**, located posterior to the maxilla and the **hamulus** of the **medial plate of the pterygoid process**, the **lateral plate of the pterygoid process**, the **scaphoid fossa**, and the **pterygoid canal of the sphenoid bone** located superior and slightly posterior to the maxilla and palatine bones (Plate 3.9). As you can see, the hard palate is made up of many bony parts. The hard palate is also the structure typically subject to facial clefting birth defects (see Branchial arches at the beginning of Chapter 3).

On the internal aspect of the skull, the temporal bone includes the **squamous part** and the **petrous part** ("stony" or "hard"). This petrous part includes the **groove for the greater petrosal branch of the facial nerve**, the **groove for the lesser petrosal nerve**, the **tegmen tympani** (portion of floor of middle cranial fossa), the **groove for the sigmoid sinus**, the **internal acoustic meatus** for the passage of the **facial nerve (CN VII)** and the **vestibulocochlear nerve (CN VIII)**. Between the temporal bone and the **occipital bone** lies the **jugular foramen** for the passage of the **glossopharyngeal nerve (CN IX)**, **vagus nerve (CN X)**, **accessory nerve (CN XI)**, and **sigmoid sinus** (Plates 3.11 and 3.12).

3.2.2 Mandible

The **mandible** is made up of the **mandibular body** (curved anterior portion) which contains the **alveolar process** with the tooth sockets, **mental protuberance (chin)**, **mental foramen** for passage of nerves and vessels, and **inferior border**; and the **mandibular ramus** (lateral, toothless portion) joined to the mandibular body by the **mandibular angle**. The mandibular ramus contains the **mandibular notch** separating the **coronoid process** (for muscle attachment) and the **condylar process**, which includes the **mandibular neck** and the **mandibular head** that articulates with the temporal bone through the **temporomandibular joint** (Plate 3.28). The chin may be unique to anatomically modern humans—although scientists continue to debate whether our close fossil relatives had chins—and it is still not clear why exactly chins evolved. The temporomandibular joint includes a **superior synovial cavity**, which is important because it allows the **articular disc** to smoothly glide onto the articular tubercle of the temporal bone (protrusion and retrusion) and an **inferior synovial cavity** that allows the mandibular head to perform hinge-like opening and closing motions.

Osteoarthritis (a disease characterized by inflammation of the joints) of the temporomandibular joint is among the most common oral health diseases. Its risk and severity increase with age due to use and misuse including stress-related jaw clenching (which is why it's important to relax your jaw even when studying and taking exams).

The **capsule of the temporomandibular joint** is loose and its lateral surface is reinforced by the **temporomandibular ligament**. On the **inner aspect of the mandible** lie the **digastric fossa** for the anterior digastric muscle, **mylohyoid line** for the mylohyoid muscle, **lingula** for the attachment of the **sphenomandibular ligament** and **submandibular fossa**, **mylohyoid groove** for the mylohyoid nerve and vessels, and **mandibular foramen** for the inferior alveolar nerve and vessels (Plates 3.8b and 3.28). These neurovascular structures pass distally in the **mandibular canal** and provide innervation to the **mandibular teeth**. Within the mandibular canal, the mental nerve branches from the inferior alveolar nerve and exits through the **mental foramen** to reach the outer surface of the mandible and innervate the chin and lower lip.

3.2.3 Neck

The **hyoid body**, **greater horn**, and **lesser horn** collectively form the small **hyoid bone** (also known as "upsilon," or "U-shaped") that lies at the angle between the floor of the mouth and the superior end of the neck. The hyoid bone is connected by the **thyrohyoid membrane** to the **thyroid cartilage** and by the **stylohyoid ligament** to the **styloid process** of the temporal bone (Plate 3.13). The **laryngeal prominence** lies on the anterior surface of the **thyroid laminae**, which are connected to the **cricoid** ("ring-like") **cartilage** by the **cricothyroid ligament**.

The **larynx** is made up of the **epiglottic cartilage** and a **stalk** that attaches it to the **thyroid laminae**, the **thyrohyoid membrane**, the **thyroid cartilage**, the **cricoid cartilage**, and the **arytenoid cartilages** (Plates 3.43 and 3.44). The epiglottic cartilage elevates during swallowing and thereby prevents food from entering the trachea; this mechanism fails when you eat and talk simultaneously. The arytenoid ("jug-like") cartilages articulate with the **cricoid cartilage** via a synovial joint and have a **muscular process** for the insertion of the laryngeal muscles and a **vocal process** for the attachment of the **vocal ligament** running from each arytenoid cartilage to the thyroid cartilage. The thyroid ("shield-like") cartilage includes two **thyroid laminae**, each including an **inferior horn** that articulates with the cricoid cartilage—**cricothyroid joint**—and a **superior horn**. The cricoid cartilage has a broad posterior **cricoid lamina** and a thinner **cricoid arch**.

Inferiorly to the cricoid cartilage lies the **trachea**, which includes the **tracheal rings**, anteriorly to which lies the **isthmus of the thyroid gland**.

Scientists continue to debate whether our closer fossil relatives, such as Neanderthals, could speak, based on what we know about the shapes and position of their laryngeal cartilages and hyoid bone. Certainly, these structures are anatomically very different between humans and our closest living relatives, chimpanzees and bonobos, and they probably evolved relatively rapidly in the lineage leading to modern humans.

The cervical (neck) vertebrae will be discussed with the remaining vertebrae in Section 6.1.

3.3 Head and Neck Neurovascular System

Before passing to the description of the head and neck neurovascular system, we should mention the scalene muscles because they are major landmarks of this region (Box 3.10). This group of muscles, which includes the **anterior scalene**, **middle scalene**, and **posterior scalene** are innervated exclusively by cervical nerves and therefore are not true head muscles and will not be described in detail in Section 3.4. They are instead **lateral vertebral muscles** covered by the **prevertebral fascia**, which also covers the **prevertebral muscles** (**longus colli** and **longus capitis**). Another important landmark is the **carotid sheath**, which contains four neurovascular structures: the **common carotid artery**, **internal carotid artery**, **internal jugular vein**, described in Section 3.3.1, and the **vagus nerve**, described in Section 3.3.2.

BOX 3.10   EASY WAYS TO REMEMBER: HYOGLOSSUS AND ANTERIOR SCALENE AS MAJOR LANDMARKS FOR NECK VESSELS AND NERVES

The neck is a difficult structure to study for most students (and doctors as well, at least some of whom refer to the neck as "tiger country"). However, there are rules that make this region much easier to learn, such as the "4 superficial vs. 2 deep" rule. Using the **hyoglossus muscle** and **anterior scalene muscle** as landmarks, remember that each of these two muscles has four structures passing superficially and two structures passing deep to it, and that one artery, but not the corresponding vein, is always part of the two deep structures. Thus, the **hypoglossal nerve CN XII**, **facial vein**, **lingual vein**, and **facial artery** are superficial to the hyoglossus muscle, while the **glossopharyngeal nerve (CN IX)** and **lingual artery** are deep to this muscle. Similarly, the **subclavian vein**, **phrenic nerve**, **suprascapular artery**, and **transverse cervical artery** are superficial to the anterior scalene muscle, while the **subclavian artery** and **roots of the branchial plexus** are deep to this muscle, passing between it and the **middle scalene muscle**, in the **interscalene triangle**. Moreover, the anterior scalene muscle is also the major landmark for the root of neck; in particular, it defines the three parts of the subclavian artery, as will be seen in Section 3.3.2.1.

3.3.1 Head and Neck Nerves

BOX 3.11   NERVE-TARGET INTERACTIONS AND NEUROGENESIS

As the **CNS** matures, the number of neurons in each part becomes precisely matched to the number of targets. This comes about through a two-step process. First, the embryo generates an excess of neurons. Later, as their axons seek targets, which include other neurons in the CNS or peripheral targets such as muscles, cells that are unable to successfully establish stable synaptic contacts die off (**apoptosis**). This principle was first demonstrated for motor neurons in the spinal cord of chick embryos, for instance: The number of ventral **motor neurons** produced during the first week of incubation exceeds by 40%–50% the number present at later stages. The same process occurs in many parts of the central and peripheral nervous systems (PNS) and also in the **retina**.

The number of cells lost is greater in the cervical and thoracic regions than in the limb-innervating brachial and lumbosacral regions. This inequality could be due to genetic programming within the spinal cord or might instead be correlated with the presence of fewer muscle fibers in the neck and trunk regions compared to the massive **limb muscles**. To answer this question, the German–American neuro-embryologist

Viktor Hamburger (formerly a student of Hans Spemann, who discovered embryonic induction) modified chicken embryos by either removing the wing bud or grafting an extra limb bud immediately adjacent to the normal wing. The results revealed that the survival of motor neurons in the CNS is dependent on the availability of target tissue. At the same time, an Italian researcher named Rita Levi-Montalcini performed similar experiments focused on the sensory ganglia. She found comparable outcomes, but discovered that the proliferation as well as survival of neurons was affected by changes in peripheral target tissues.

What does "availability" of peripheral targets actually mean? Is direct contact required or is there some substance produced by limb tissues— by muscles for motor nerves—that prevents these neurons from dying? The answer is mixed. Noting that there were some tumors that became heavily innervated, an assistant hired by Professor Hamburger placed a piece of this sarcoma tissue beside the somites of a chick embryo. The results were astounding. Nearby neurons that normally would have died in **peripheral ganglia—sensory neurons** in **spinal ganglia** and **sympathetic neurons** in ganglia located ventral to vertebrae—instead survived. But the most revolutionary discovery was that the same was true in distant ganglia whose axons did *not* contact the tumor. Clearly, something produced by the tumor was rescuing neurons from death. Indeed, placing tumor tissue outside the embryo, on the blood vessels of the chorio-allantoic membrane had the same effect.

This substance was found to be a peptide that was named **nerve growth factor**, which *was the first growth factor ever discovered and characterized*, and for her research Levi-Montalcini received a Nobel Prize in 1986. But what about motor neurons? The nerve growth factors had no rescuing affect on this population, yet they are also target-tissue dependent. Here the story is more complex; motor neurons are a heterogeneous population of neurons and several growth factors (and their receptors) have been found that alter the survival of motor neuron subsets. In addition, both "survival" and "killing" factors have been identified in embryonic muscles, suggesting that developing motor neurons critically regulate their response capabilities. All of these growth factors are taken up by the terminal growth cones or presynaptic swellings on axons and then transported via retrograde flow to the cell bodies. Not surprisingly, disruption of this transport process also compromises the survival of neurons, both in the embryo and adult.

BOX 3.12   NOTES ON THE DEVELOPMENT AND ANATOMY OF THE PERIPHERAL NERVOUS SYSTEM

The **PNS** encompasses all neurons that have cell bodies and/or axonal projections located outside of the brain and spinal cord. This includes both motor (efferent) components that convey signals away from the **CNS** and sensory (afferent) projections bringing signals into the CNS (note that a nerve is a collection of nerve fibers and associated Schwann cells and connective tissues). **Sensory (afferent) neurons** are located in segmentally arranged aggregations called **spinal ganglia** (also called dorsal root ganglia) beside the spinal cord and in **cranial sensory ganglia** located beside the hindbrain. Sensory neurons serve both somatic (e.g., body wall, voluntary muscles, and tendons) and visceral (e.g., smooth muscles associated with gut and blood vessel walls) targets (note that a ganglion is an aggregation of peripheral neurons, which can be either sensory or motor). **Somatic efferent neurons** are located in the motor columns of the spinal cord and brainstem. Axons from somatic efferent neurons project directly to their target skeletal muscles. These nerve fibers exit the spinal cord as **ventral roots** that, once joined by sensory fibers, make up the segmentally arranged **spinal nerves** (see Section 6.2).

The **autonomic nervous system** (Plate 3.20) is a fully separate motor system that controls involuntary peripheral functions such as vascular tone, constriction and dilation of the pupil, gut peristalsis, and glandular secretions, for example, lacrimal, salivary, and sweat glands. It has three subdivisions: **sympathetic system**, **parasympathetic system**, and **enteric system**. The autonomic motor system, sometimes referred to as the involuntary or **visceral efferent system**, differs anatomically from the somatic motor system in that at least two autonomic motor neurons are arranged in series to link the CNS to terminal peripheral targets. Cell bodies of the first neurons are located in the brainstem or intermediate motor column of the spinal cord. Peripheral projections from these neurons exit the spinal cord via ventral roots and are called **preganglionic axons**. They synapse with the second neurons located in peripheral **autonomic ganglia**. **Postganglionic axons** project from these neurons to target tissues. In general, the **sympathetic nervous system** facilitates the body's response to fright and stress, for example, increased heart rate and reduced peripheral circulation due to constriction of perivascular smooth muscles. The term "wide eyed with fright" describes the action of sympathetic innervation on dilator muscles of the **iris**. Many **sympathetic ganglia** are located ventrolateral to thoracic and lumbar vertebrae and others are scattered throughout the thorax and head.

Most sensory and all autonomic neurons whose cell bodies are located outside the brain and spinal cord are derived from **neural crest cells**. After breaking away from the roof of the neural tube shortly after it closes, and on completing the epithelial-to-mesenchyme transformation, crest cells migrate to multiple peripheral regions. The derivatives of the neural crest are heterogeneous and include peripheral sensory and autonomic neurons plus associated glia, which support and nourish the neurons and **Schwann cells** (described below). Crest cells are also the source of all peripheral pigment cells (i.e., those outside the eye and brain), and many connective tissues in the head, as explained above. Trunk crest cells detach from the roof of the neural tube in a cranial-to-caudal sequence and then begin moving peripherally, encountering the dorso-medial margin of the **somite** just as **sclerotome** is separating from the **dermamyotome** (the sclerotome and dermamyotome are the derivatives of the somites, which are described in Section 2.1). Neural crest cells are generated as a continuous mesenchymal population, but many of their derivatives, for example, **spinal ganglia** become segmentally arranged. This transformation occurs because the caudal half of each somite inhibits crest cell movements. As a result, crest cells become segmentally restricted to the cranial half of each somite. Within the somite, some crest cells stop moving and aggregate; these are the precursors of sensory neurons and glia of the spinal ganglia. The remaining crest cells move ventrally through the cleft between the sclerotome and myotome and emerge near to the dorsal aortae (embryonic structures associated with the brachial arches; see "branchial arches" at the beginning of Chapter 3). Here, many of the crest cells aggregate to form precursors of the **sympathetic ganglia**. Other crest cells move to sites where major arteries branch ventrally from the aorta and form thoracic and abdominal sympathetic ganglia. While passing through somites, some crest cells encounter motor axons that have grown out as ventral roots from the spinal cord; these cells stop moving and bind to nerve fibers, becoming Schwann cells. Later, they will separate and envelop individual axons and produce insulating layers of myelin. A final population of emigrating crest cells does not enter the somite, but moves laterally between the dermamyotome and overlying surface ectoderm. These cells will form peripheral **pigment cells**, the **melanocytes**.

The **parasympathetic system** maintains homeostasis in the body, often countering the effects of sympathetic function. It is primarily associated with internal organs and does not have the extensive peripheral distribution of sympathetic fibers. Like the sympathetic system, the parasympathetic system has two (or more) neurons between the CNS and its targets. However, there are two significant structural differences: (1) the first neurons are all located in the brain or spinal cord, not the thoracolumbar region and (2) the second neurons are usually located in the wall of the target organ, not in ganglia.

The developmental relationships between the head and neck nerves, muscles, and structures such as the larynx, pharynx, and the branchial (pharyngeal) pouches are illustrated in Plate 3.2. The best way to learn the nerves of the head and neck is to take a holistic approach: First, look at the entire region and notice how all the nerves relate to each other and to the surrounding structures (Plates 3.21 through 3.25); second, try to understand the overall configuration of a nerve and of its major branches using the illustrations of individual nerves; third and finally, study each of these branches in detail and their relationships with the associated regional structures in Section 3.5. Most students find it easier to begin with a big-picture view and then move to the specifics, rather than the other way around.

3.3.1.1 Olfactory Nerve CN I

The **olfactory nerve** (**CN I**) (Plate 3.39) is actually not a true peripheral nerve, but an outgrowth of the brain itself. The neurons that lie in the olfactory mucosa and send their axons through the holes of the cribriform plate are the real peripheral nerves of the olfactory system.

3.3.1.2 Optic Nerve CN II

The bilayered retina is an outgrowth of the brain and ganglion cells located in this tissue send axons beneath the optic stalk and into the brain as the **optic nerve** (**CN II**) (Plate 3.14; see also Section 3.5.7). The glial cells that support the olfactory and optic nerves are of the type that surrounds CNS structures and therefore are not damaged by demyelinating diseases that target Schwann cells (e.g., multiple sclerosis).

3.3.1.3 Oculomotor Nerve CN III

The **oculomotor nerve (CN III)** (Plate 3.14) innervates five ocular muscles (see Section 3.4.5 for details about the extraocular muscles). The **superior division of the oculomotor nerve** logically innervates the two more superior muscles—**levator palpebrae superioris** that elevates the eyelid and the **superior rectus** that elevates the eye. The **inferior division of the oculomotor nerve** innervates the **inferior rectus** that depresses the eye, the **medial rectus** that adducts the eye, and the **inferior oblique** that elevates and abducts the eye because it runs to the posterolateral portion of the inferior surface of the eye. The oculomotor nerve carries with it the axons of parasympathetic neurons located near its somatic motor neurons. These axons leave the oculomotor nerve to synapse in the ciliary ganglion in the orbit. Postganglionic fibers enter the eyeball and travel to the iris, where they cause contraction of the sphincter pupillae muscle, thereby constricting the pupil.

3.3.1.4 Trochlear Nerve CN IV

The **trochlear nerve (CN IV)** innervates the **superior oblique muscle**, which depresses and abducts the eye (Plates 3.14 and 3.33). It runs anteriorly from the brainstem and joins CN III and CN VI before all three nerves enter the orbit through the superior orbital fissure (see Section 3.5.7 for descriptions and actions of the extraocular muscles).

3.3.1.5 Trigeminal Nerve CN V

The **skin** of the face receives sensory innervation from the **trigeminal nerve (CN V)** (Plate 3.15). This nerve has three branches (divisions): The **ophthalmic nerve (CN V_1)** innervates the skin of the **forehead, upper eyelids**, and dorsum of the **nose**, and gives rise to the **supraorbital nerve** that passes through the **supraorbital foramen**; the **maxillary nerve (CN V_2)** innervates the **lower eyelids, cheek**, and upper lip. The trigeminal nerve gives rise to the **infraorbital nerve** that emerges from the **infraorbital foramen**. The **mandibular nerve (CN V_3)** innervates the skin of the **lower face**—including the **chin** and **lower lip**—and part of the side of the head and gives rise to the **mental nerve** that emerges with the **mental artery** and **mental vein** from the **mental foramen**. The back of the head and the area around the ear receive sensory innervation from **spinal cervical nerves 2 and 3** (Section 3.3.1.12).

The **ophthalmic nerve (CN V$_1$)** gives rise to three main nerves that lie in the orbital region: the **frontal nerve**, which runs anteriorly and branches into the **supratrochlear nerve** and the **supraorbital nerve**; the **lacrimal nerve**, which runs anterolaterally to reach the lacrimal gland and lateral eyelid; and the nasociliary nerve, which runs inferomedially and sends branches through the **ciliary ganglion** (without synapsing) to the eyeball (short ciliary nerves) and directly to the eyeball (**long ciliary nerves**). The nasociliary nerve also gives rise to the **anterior and posterior ethmoidal nerves** (Plates 3.15 and 3.33). Both of these innervate the ethmoid sinuses, but the anterior ethmoid nerve also runs through the nasal cavity and to the dorsum of the nose as the external nasal nerve. None of these nerves innervate muscles.

The **maxillary nerve (CN V$_2$)** also has many sensory branches. The **nasopalatine nerve** is a branch of CN V$_2$ that lies in the mucosa of the nasal septum and runs with the **sphenopalatine artery** from the sphenopalatine foramen to the **incisive canal** and supplies the nasal septum and the mucosa that covers the anterior hard palate (Plate 3.39). In the lateral wall of the **nasal cavity**, the **lesser and greater palatine nerves** descend from the maxillary nerve, running through the **pterygopalatine ganglion** into the **greater palatine canal**, and pass through the lesser and greater palatine foramina, respectively. Parasympathetic fibers run with the maxillary nerve, having originated as branches of the facial nerve. These facial branches run first as the **greater petrosal branch of the facial nerve**, then as the **nerve of the pterygoid canal (Vidian nerve)**, and then synapse in the pterygopalatine ganglion (see Section 3.3.1.7). Postganglionic axons that arise in this ganglion are distributed with branches of the maxillary nerve (Plate 3.16). The nerves of the pterygopalatine ganglion stimulate the lacrimal gland (crying) and also secretions from the mucosa of the nasal cavity, paranasal sinuses, roof of the mouth, soft palate, and nasopharynx.

The **mandibular nerve (CN V$_3$)** gives off the **inferior alveolar nerve**, which passes through the **mandibular foramen**, runs with the **inferior alveolar vessels**, and then passes distally into the **mandibular canal**. The inferior alveolar nerve subsequently innervates the **mandibular teeth** before it gives off a branch called the mental nerve, which passes through the **mental foramen** visible on the outer surface of the mandible to innervate the chin and lower lip. Note that the **mylohyoid nerve** is a branch of the inferior alveolar nerve that does not go through the mandibular canal; it provides motor innervation to the **mylohyoid muscle**, so it runs instead through the **mylohyoid groove** on the inner side of the mandible. The **lingual nerve** is also a branch of the mandibular nerve (CN V$_3$): It emerges between the lateral and medial pterygoid muscles just anterior to the inferior alveolar nerve. Here, the lingual nerve passes medial to the 3rd **mandibular molar tooth** and innervates the mucosa of the anterior two-thirds of the **tongue** and floor of the **oral cavity**. This explains why during an invasive dental procedure, when a **mandibular nerve block** is produced by injecting a **dental anesthesia** agent into the infratemporal fossa, the drug will anesthetize not only the mandibular teeth (inferior alveolar nerve) but also the lower lip and chin (mental nerve) and the tongue (lingual nerve). The lingual nerve carries branches supplying the mucosa of the anterior two-thirds of the tongue with taste fibers and general sensation, and sends two short branches to the **submandibular ganglion**. Also note that the **chorda tympani** joins the posterior side of the lingual nerve. Other branches of the mandibular nerve (CN V$_3$) that lie in the temporal region are the **buccal nerve** and the **auriculotemporal nerve**. The auriculotemporal nerve passes between the head of the **mandible** and the **external acoustic meatus** and crosses the **zygomatic process** of the **temporal bone** to innervate the skin of the anterior side of the ear and temporal region. The auriculotemporal nerve carries postganglionic parasympathetic nerve fibers from the **otic ganglion** to the **parotid gland**. The **buccal nerve** emerges from deep to the masseter muscle to innervate the skin of the cheek, and to pierce the buccinator to provide sensory innervation to the **mucosa of the cheek**.

The mandibular nerve is the only division of the trigeminal nerve that carries motor neurons in addition to sensory neurons. The muscles that it innervates are mostly muscles related to the jaw and are all derived from the 1st branchial arch. They include the four major muscles of mastication or chewing (**masseter, temporalis, lateral pterygoid** and **medial pterygoid**) as well as the **anterior digastric**, the **mylohyoid**, the **tensor veli palatini**, and the **tensor tympani** (Section 3.4.1).

3.3.1.6 Abducens Nerve CN VI

The **abducens nerve** (**CN VI**) (Plate 3.14) innervates the **lateral rectus muscle**, which abducts the eye. It enters the orbit through the superior orbital fissure with CN III and CN IV (see Section 3.5.7 for descriptions and actions of the extraocular muscles).

3.3.1.7 Facial Nerve CN VII

The **muscles of facial expression** are innervated by the **facial nerve (CN VII)** (Plates 3.16, 3.26, and 3.27). Just anterior to the ear lobe, a few branches of the facial nerve emerge from the **stylomastoid foramen**, including the branches that form the parotid plexus within the **parotid gland**. The parotid plexus of the facial nerve gives off the **temporal branch** crossing the zygomatic arch; the **zygomatic branch** crossing the zygomatic bone; the **buccal branches** crossing the superficial surface of the masseter muscle; the **mandibular (or marginal mandibular)** branch that parallels the inferior margin of the mandible; and the **cervical branch** that crosses the angle of the mandible to enter the neck. In human anatomy, these are usually considered to be *branches* of the facial nerve, and not discrete nerves themselves. For instance, the **buccal branch of the facial nerve** crosses the superficial surface of the masseter muscle to provide motor innervation to the buccinator muscle, while the **buccal nerve** of the trigeminal mandibular division (CN V_3) emerges from deep to the masseter muscle to innervate the skin of the cheek, and to pierce the buccinator to provide sensory innervation to the cheek mucosa. Other branches of the facial nerve are the **posterior auricular nerve**, which innervates some facial expression muscles, and the nerves to the **stylohyoid muscle** and to the **posterior digastric muscle**, which innervate the stylohyoid and posterior digastric muscles, respectively.

Other branches of the facial nerve contain special sensory and parasympathetic fibers. The **chorda tympani** passes between the incus and malleus, near the **tympanic membrane** (Plate 3.45a and c), that emerges from the **temporal bone** just posterior to the **temporo-mandibular joint**, and joins the lingual nerve of the trigeminal nerve complex. The special sensory nerves carry taste sensation from the anterior two-thirds of the tongue. The parasympathetic fibers synapse in the **submandibular ganglion**. Postganglionic axons innervate the **submandibular salivary gland** and **sublingual salivary gland**. Another branch of the facial nerve, the **greater petrosal branch**, leaves the temporal bone on the petrosal surface, and joins the **deep petrosal nerve** to form the **nerve of the pterygoid canal**. The nerve of the pterygoid canal enters the pterygopalatine fossa, where the parasympathetic fibers synapse in the **pterygopalatine**

ganglion. The deep petrosal nerve carries postganglionic sympathetic fibers that will pass through the pterygopalatine ganglion to be distributed with the sensory fibers of the maxillary division of the trigeminal nerve (Plate 3.39b). As will be explained in Section 3.3.1.5, the ganglion stimulates the lacrimal gland and secretions from the mucosa of the nasal cavity, paranasal sinuses, roof of the mouth, soft palate, and nasopharynx.

3.3.1.8 Vestibulocochlear Nerve CN VIII

As noted above, the **vestibulocochlear nerve** (**CN VIII**) (Plate 3.45a) enters the internal acoustic meatus together with the facial nerve and, as its name indicates, mainly innervates the **vestibulocochlear organ** of the inner ear. There it splits into two divisions, one going to the cochlea (hearing) and one going to the vestibular system (equilibrium) (see Section 3.5.13 for information about these two senses).

3.3.1.9 Glossopharyngeal Nerve CN IX

The general sensory fibers of the **glossopharyngeal nerve** (**CN IX**) (Plate 3.17) innervate the mucosa of the tympanic cavity, the mucosa of the posterior one-third of the tongue and the posterior wall of the oropharynx. The glossopharyngeal nerve also carries taste fibers from the posterior one-third of the tongue. It innervates a single pharyngeal muscle, the **stylopharyngeus**.

3.3.1.10 Vagus Nerve CN X

BOX 3.13   DESCENT OF THE HEART AND ASYMMETRIC PATHWAYS OF THE RECURRENT LARYNGEAL NERVE

During embryonic development, **aortic arches** within each branchial arch develop, and pass from the heart, located ventrally, to the paired dorsal aortae located dorsal to the pharynx. As cardiac looping is underway (an early step in transformation of the heart tube to the adult chambered heart), the heart begins to shift caudally and additional aortic arches are formed. This process accompanies elongation of the **pharynx**, **esophagus**, and **trachea**, and is accentuated by a cranial elongation of dorsal axial structures, which is driven by growth of **cervical vertebrae** (early embryos have no neck, as can be seen in Plate 2.1). See Section 3.3.2 and Box 3.15 for a detailed description of normal arterial and cardiovascular development (Plate 3.4).

The multiple branches of the **vagus nerve** (CN X) (Plate 3.18) enter the 4th and 6th branchial arches and wrap around the aortic arches in each. The subsequent descent of the heart pulls the 4th and 6th aortic arches and **subclavian arteries** caudally. As these vessels shift caudally, branches of the vagus nerve called the **recurrent laryngeal nerves** are snagged behind the 6th aortic arch and become greatly elongated.

When the dorsal part of the right 6th aortic arch degenerates, the snagged vagus nerve on this side slips cranially until it is caught by the right 4th arch, which remains as the connection to the right subclavian artery. On the left, however, the recurrent laryngeal nerve remains caught behind the **ductus arteriosus**, and in the adult is found wrapped behind the **ligamentum arteriosum**. The recurrent laryngeal nerves change name at the cricothyroid joint (between the cricoid and thyroid cartilages), to inferior laryngeal nerves, which innervate all intrinsic laryngeal muscles. Because they pass close to the **thyroid gland**, damage to the recurrent laryngeal nerves during **thyroidectomy** is a significant concern. **Lymph nodes** located beside these nerves are also frequent sites of **metastasis** from **lung tumors**. This condition causes the nodes to enlarge and the resulting pressure on the recurrent laryngeal nerves can cause laryngeal paralysis, clinically recognized by difficulty in speaking clearly.

The **vagus nerve** (**CN X**) (Plate 3.18) has a complex overall configuration, in accord with the English translation of its Latin name as "vagrant" or "wanderer." It innervates many different structures including all the true **pharyngeal muscles** (including the **cricothyroid muscle**, which is truly an extension of the **inferior pharyngeal constrictor muscle**, and not a true laryngeal muscle) except the **stylopharyngeus** (innervated by the **glossopharyngeal nerve**, or **CN IX**). The vagus nerve also innervates all the true **laryngeal muscles**, the **trachea**, and the **heart**, as well as the **digestive tract**. In addition to the spatially close relationships of the heart with the digestive tract, these two structures are closely connected in a developmental sense because the heart is formed during the process of gut tube closure. Although the heart is a single, median ventral structure, cardiac primordia are established bilaterally before folding occurs (Figure 3.4). As folding of the gut tube and body wall

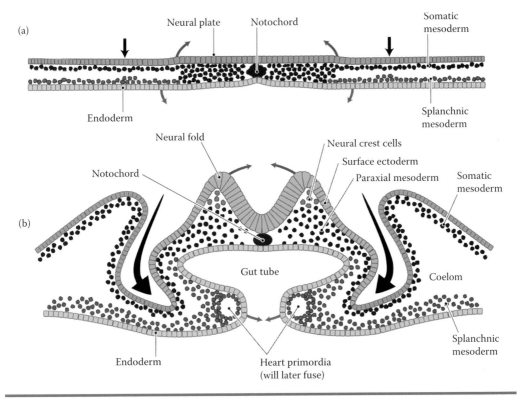

Figure 3.4 **(a and b) Schematic transverse views of the early embryo (cut off from all extra-embryonic tissues) showing the folding of the superficial (blue) layer called to form the neural tube and the lower (yellow) layer to form the gut tube. Note the separation between somatic (red) and splanchnic (orange) layers of mesoderm. Black arrows show the deep body folds that will undercut the entire early embryo and creating a cylindrical body that is separated from extra-embryonic tissues except at the umbilical stalk.**

occur, these cardiac primordia are carried to the ventral midline, beneath the gut, and subsequently fuse together to form the single cardiac tube.

A major branch of the vagus nerve is the **superior laryngeal nerve**, which in turn divides into the **internal laryngeal nerve** (sensory and parasympathetic) and the **external laryngeal nerve** (motor to the crico-thyroid and inferior pharyngeal constrictor muscles). It should be noted that all intrinsic (true) laryngeal muscles are innervated by the **left and right inferior laryngeal nerves**, which are continuous with the **left and right recurrent laryngeal nerves** and provide sensory and parasympathetic innervation to the larynx below the **vocal folds** and parasympathetic, efferent, and afferent innervation to the upper esophagus, and the trachea. The left and right recurrent laryngeal nerves branch from the left and right vagus nerves at the level of the **aortic arch** and of the **right subclavian artery**, respectively, to loop under these structures before ascending to the laryngeal region. The internal laryngeal nerve changes its name to inferior laryngeal nerve when it passes inferiorly to the level of the **glottis**. The inferior laryngeal nerve then changes name to recurrent laryngeal nerve when it passes inferiorly to the level of the cricothyroid joint (Plate 3.42). It is very easy to identify the vagus nerve on the neck region because is lies within the carotid sheath, between the **internal jugular** vein and the **common carotid artery** (Plate 3.24). The **pharyngeal plexus of the vagus nerve** joins the **pharyngeal plexus of nerves** which is located on the posterolateral surface of the pharynx (Plates 3.17 and 3.20).

3.3.1.11 Accessory Nerve CN XI

The **accessory nerve (CN XI)** (Plate 3.19) is said to have two roots: a **cranial root** that joins the **vagus nerve (CN X)** and that provides motor innervation to the true **laryngeal muscles** via the **inferior laryngeal nerve**; and a **spinal root** that provides motor innervation to the sternocleidomastoid and the trapezius. However, the cranial root is now often considered to be part of the vagus nerve. Branches of **spinal nerves C3 and C4** join the accessory nerve in the **posterior cervical triangle** (Plate 3.24) to provide sensory innervation to the sternocleidomastoid and the trapezius.

BOX 3.14   THE STRANGE CASE OF CRANIAL NERVE XI

The **accessory nerve** (CN IX) emerges from the floor of the brainstem close to the vagus nerve and innervates the **trapezius** and **sternocleidomastoid** muscles. This innervation pattern appears to violate the rule that nerve–muscle relationships reflect segmental embryonic history: How could an upper limb-moving muscle such as the trapezius be developmentally related to branchial arches and brainstem somatic motor nerves? The answer lies is our evolutionary history: Fish have no neck! Parts of the pectoral girdle skeleton that support the pectoral fin in many fish articulate with roofing bones of their skull. In cartilaginous fish such as sharks, the precursor of the amniote trapezius and sternocleidomastoid muscles attaches to both the brachial arch skeletal structures and the pectoral girdle. During vertebrate evolution, cervical vertebrae were added and caudal parts of the skull roof were lost, creating a gap between head/ skull/branchial arches and the pectoral girdle. In amniotes, the progenitors of the trapezius and sternocleidomastoid muscles initially develop in association with the caudal branchial arch tissue and the caudal part expands caudally at later stages to attach onto postcranial bones (such as the scapula, in humans) gaining sensory innervation from some cervical nerves (C3 and C4) (see Section 3.4.3.3).

3.3.1.12 Hypoglossal Nerve CN XII and Cervical Nerves

Together with the **hypoglossal nerve (CN XII)** (Plate 3.19), the **cervical nerves** provide motor innervation for all neck and tongue muscles that are derived from **somites**, but *not* from the **branchial arches**—that is, it does not innervate **branchiomeric muscles**, which are the head muscles in an evolutionary and developmental context. The **infrahyoid muscles** are closely related developmentally and evolutionarily to the **tongue muscles** because they are somitic muscles formed by myoblasts that migrated ventrally, around the pharynx, then cranially to be *secondarily* incorporated into the head and neck. This migration is clearly seen during development, where these muscles migrate via the **hypoglossal cord (chorda hypobranchialis)** (Plate 3.2). This somitic origin is why the infrahyoid and tongue muscles are both part of the so-called hypobranchial group of muscles. Therefore, it is not surprising that axons of the spinal nerve C1 are carried by the hypoglossal nerve to innervate a **tongue muscle** (the **geniohyoid**) and an **infrahyoid muscle** (the **thyrohyoid**, via the **nerve to the thyroid muscle**). The other three infrahyoid muscles (**omohyoid, sternothyroid,** and **sternohyoid**) receive motor innervation from the **ansa cervicalis**, which is formed by the first, second, and third cervical nerves. The other three true tongue muscles (**genioglossus, hyoglossus,** and **styloglossus**) receive motor innervation from the hypoglossal nerve (CN XII).

The hypoglossal nerve is easy to identify because it runs posteroanteriorly first, as one of the four major structures passing superficial to the **hyoglossus muscle** (see Box 3.10), and then deep to the **mylohyoid muscle** (Plate 3.41). The **superior root of the ansa cervicalis** travels with the hypoglossal nerve and is mainly composed of fibers from C1, whereas the inferior root of the ansa cervicalis is mainly composed of fibers of C2 and

C3, and passes around the carotid sheath to join the superior root, thence the name *"ansa"* (handle).

The cervical nerves also provide sensory innervation for two true (i.e., branchiomeric) head muscles via the **cervical plexus**: nerve C2 and C3 to the sternocleidomastoid and nerves C3 and C4 to the trapezius. The cervical nerves also provide cutaneous innervation to the skin of the neck and posterior region of the head. The **lesser occipital nerve**, **great auricular nerve**, **transverse cervical nerve**, and the **supraclavicular nerve** are branches of the cervical plexus that enter the **superficial fascia** posteriorly to the **sternocleidomastoid** to supply distinct areas: the **scalp**, the skin of the lower part of the ear and **angle of mandible** and **mastoid process**, the skin of the **anterior triangle of the neck**, and the skin of the superior region of the **shoulder**, respectively. In contrast, the dorsal rami of the cervical spinal nerves mainly pierce the trapezius and go superficially to innervate the back of the head and neck. One of these nerves is the **greater occipital nerve**, hence the name "lesser occipital nerve" for the nerve coming from the cervical plexus.

The **sympathetic trunk**, the **superior, middle, and inferior cervical sympathetic ganglia** (and **cervicothoracic ganglion—or stellate ganglion**—a ganglion formed by the inferior cervical ganglion and the **1st thoracic ganglion**) are connected by the **gray rami communicantes** to the cervical spinal nerves. The **internal carotid nerve** runs from the superior cervical ganglion to the internal carotid artery, as its name indicates (Plate 3.20).

3.3.2 Head and Neck Blood Vessels

3.3.2.1 Head and Neck Arteries

BOX 3.15   BASIC CONCEPTS IN ARTERIAL DEVELOPMENT

In the adult human, the number of smooth muscle layers is greater in **arteries** than in **veins**, and in both cases it is proportional to the diameter of the vessel. Capillaries lack smooth muscle, but are partially wrapped by **pericytes**, which also serve as smooth muscle stem cells and are activated following vascular injury. Precursors of most adult arteries are present by the time the embryo is 6–8 mm long, but the topographical relationships among organs and between regions subsequently undergo many changes. During this prolonged period of growth, the initial spatial relationships and symmetry of the arterial system become distorted, especially when organs change their relative positions. For example, the configuration of the arteries must change when the heart shifts from the level of the head to the level of the thorax, when the intestinal tract elongates and rotates, and when the testicles descend from the abdomen to the scrotum.

The arterial system is initially symmetrical—most vessels form bilaterally and are present initially on both sides (e.g., aortic arches, most of the dorsal aorta). Only a few arteries, for example, the ventral aorta, are initially singular, and these are found in the midline. In the adult, the arterial system is more asymmetrical, especially in the thorax. Consider, for example, the asymmetry of vessels that carry blood away from the heart (**pulmonary trunk**, **ascending aorta**, **coronary arteries**) and the irregular branching patterns of the **left and right subclavian arteries**. Transformation from the embryonic to the adult vascular anatomy is driven by a few basic processes: (1) *selective degeneration* of vessels, which can be unilateral or bilateral; (2) *remodeling*, which changes the relative lengths of vascular channels and also allows their branching sites to shift positions; and (3) tremendous *increase in both the number of vessels and the total surface area available* for gas and nutrient exchange with surrounding tissues, achieved in large part through the continued *formation and expansion of capillary plexuses.*

The segmentally arranged series of **aortic arches** (Plates 3.4 and 3.5), **branchial arches**, and **pharyngeal pouches** are present in all vertebrate embryos (Plate 3.2), but their adult forms have been highly modified during evolution, especially with the loss of gill-based respiration during tetrapod evolutionary history. In adult fishes, the pharyngeal pouches persist as slits between each gill, and two aortae are present: The ventral aorta carries blood to the gills, and the dorsal aorta carries blood away from the gills. During human development, a single ventral **aorta** emerges cranially from the **heart**. Arising from this aorta is a series of paired aortic arches that carry blood dorsally, around the pharynx, and into paired dorsal aortae (also called dorsal aortic roots). Each aortic arch is located within a branchial arch, which as described earlier is part of a series of mesenchymal condensations located beside and beneath the pharynx. Each aortic arch is immediately rostral to a pharyngeal pouch, which is a lateral outpocketing of pharyngeal endoderm. From these aortic arches, the paired dorsal aortae carry blood caudally, giving off multiple branches to axial structures, body wall and limbs, and viscera.

At the level of the diaphragm, the left and right dorsal aortae fuse to form a single aorta, which extends through the abdomen and branches to become paired once again. During development, both the **vitelline arteries** to the **yolk sac** and **umbilical (allantoic) arteries** to the **placenta** branch from the dorsal aortae.

Following the terminology proposed by Noden and colleagues (see the first paragraphs of of this chapter), the 4th aortic arch is the most caudal of the series; that is, there is no 5th aortic arch in mammals (Plates 3.4 and 3.5). New vessels branch from the ventral base of each 4th aortic arch and grow beside the **respiratory diverticulum** (future **trachea** and **lungs**), establishing the **left and right pulmonary arteries**. Subsequently the root of the 4th aortic arch expands, forming an enlarged aortic sac. Later, the aortic sac will become subdivided into separate systemic (4th arch) and pulmonary (the so-called "6th arch," see above) compartments. The left and right dorsal aortae between arches 3 and 4 degenerate. This degeneration effectively segregates blood going to the head via aortic arches 3 and from blood going to the trunk via aortic arches 4 and 6. The 3rd aortic arches become the **common carotid arteries** and the dorsal and ventral aortae formerly associated with aortic arches 1 and 2 remain to form the **internal carotid arteries** and **external carotid arteries**, respectively.

Why do connections form between the pulmonary arteries and the dorsal aortae? They form to meet the respiratory needs of the embryo, in which blood is cleared of $CO2$ and is oxygenated by the placenta, not the lungs. Normally, blood entering the heart would pass through the right atrium and ventricle, and then go to the lungs. However, the embryonic lungs are nonfunctional. The embryo establishes links from the pulmonary arteries to the dorsal aortae, thereby allowing oxygenated blood to flow directly into the systemic circulatory system. This link is retained only on the left side as the **ductus arteriosus** (Plates 3.4 and 3.5). The ductus arteriosus shunts >90% of blood entering the right ventricle into the aorta. It is often larger in diameter than the aorta emerging from the heart. Immediately after birth, the lungs expand, which greatly reduces resistance to blood flow through pulmonary arteries. At the same time, the smooth muscles in the wall of the ductus arteriosus contract in response to elevated oxygen concentrations and falling levels of prostaglandin PGE2 levels (PGE2 is produced by the placenta). This causes the ductus to lose patency, usually within 24 hours. Remnants of the ductus persist in the adult as the **ligamentum arteriosum**.

The best way to learn the arteries of the head and neck may be to envision them as continuations of the branches of the **aortic arch** to the different regions of the head and neck. Therefore, we begin with the three major branches of the aortic arch: (1) the **brachiocephalic trunk** that gives rise to the **right subclavian artery** and the **right common carotid artery**; (2) and (3) the **left common carotid artery** and **left subclavian artery** that branch directly from the aortic arch.

BOX 3.16 DEVELOPMENT OF THE SUBCLAVIAN ARTERIES

The right and left **dorsal aortae** are positioned immediately ventral to somites in the future neck and thoracic regions (Plates 3.4 and 3.5) and then are fused in the midline in the abdominal region. They give off a series of **intersegmental arteries** that grow dorsally between each somite, beginning under the hindbrain. The most cranial of these arteries degenerate, but the 7th intersegmental artery persists and enlarges. It gives off branches that grow into the upper limbs, thoracic body wall, and also cranially into the head immediately beside the vertebrae. These paired 7th dorsal intersegmental arteries become the proximal parts of the right and left **subclavian arteries** (Plates 3.4 and 3.5). Blood flows through the left and right 4th aortic arches into the paired dorsal aortae, then through the 7th intersegmental arteries, and finally into several branches (e.g., **axillary artery**, **internal thoracic artery**, **vertebral artery**).

This early, symmetrical pattern changes over the course of embryonic development. The first change is a *focal degeneration of the right dorsal aorta between the site where the subclavian artery branches and the site where the left and right dorsal aortae fuse together* (Plates 3.4 and 3.5). Blood entering the left 4th arch flows into the left subclavian and continues caudally into the vessel which will become the adult aorta; on the right side, blood from the 4th aortic arch goes only to the right subclavian artery and its branches.

After this asymmetrical change, the neck begins to elongate and ventrally-located structures such as the esophagus and heart shift caudally (often referred to as the "**descent of the heart**"). As this happens, the left and right subclavian arteries are "snagged" around the **1st rib**, which pulls them cranially relative to their original branching locations from the dorsal aortae (Plates 3.4 and 3.5). Because the right dorsal aorta immediately caudal to the subclavian artery is lost, the subclavian on this side is "free" to shift cranially and moves all the way up to the region of the carotid arteries. On the left, the subclavian artery also shifts cranially, but not as far, usually remaining on the arch of the aorta.

Among humans, and placental mammals in general, a great variety of branching patterns result from vascular remodeling and heart descent-related movements. In addition, several variations often occur within each species. These variations have clinical significance, for instance, in the event of surgeries associated with these vessels, including cardiac catheterizations.

In adult humans, the subclavian artery is one of the two structures (together with the **roots of the branchial plexus**) deep to the **anterior scalene muscle**, contrary to the subclavian vein, which is one of the major four structures superficial to this muscle together with the **phrenic nerve**, **suprascapular artery**, and **transverse cervical artery** (the anterior scalene muscle is a major landmark in the neck and follows the "4 superficial vs. 2 deep" rule: See Box 3.10). Because these two latter arteries are branches of the subclavian artery (which is deep to the anterior scalene muscle), they must run superficially and appear anteriorly to the anterior scalene muscle and run posteriorly and superficially to it (Plate 3.22). The **1st, 2nd, and 3rd parts of the subclavian artery** are medial, posterior, and lateral to the anterior scalene muscle, respectively. The 1st part has three branches: (1) the **vertebral artery** running superiorly between the **anterior scalene muscle** and the **longus colli muscle** and entering the transverse foramen of **vertebra C6**; (2) the **internal thoracic artery** running inferiorly to supply the anterior thoracic wall; and (3) the **thyrocervical trunk** giving rise to the suprascapular artery (described in Chapter 4) and transverse cervical artery (which supplies the **trapezius** muscle) described earlier, and to the **inferior thyroid artery**, which passes posterior to the **cervical sympathetic trunk** toward the **thyroid gland** and gives rise to the **ascending cervical artery**. The 2nd part of the subclavian artery has one branch: the **costocervical trunk**, which gives rise to the **deep cervical artery** and the **supreme intercostal artery** (which gives rise to the **posterior intercostal arteries 1 and 2**). Lastly, the third part of the subclavian artery gives off a single branch—the **dorsal scapular artery**—before changing its name to the **axillary artery** at the level of the **1st rib** (both the dorsal scapular and the axillary arteries are described in Chapter 4). It should be noted that the thoracic duct lies in this region of the root of the neck, draining into the junction of the **left subclavian vein** and the **left internal jugular vein**.

The **common carotid artery** bifurcates into the **external carotid artery** and the **internal carotid artery** and has no other major branches. The internal carotid artery also has no branches in the neck being a major supplier to the brain. In contrast, the external carotid has six branches (Box 3.18) before it divides, superiorly, into the **maxillary artery** and **superficial temporal artery**.

BOX 3.17   EASY WAYS TO REMEMBER: VESSELS AND NERVES OF THE ROOT OF THE NECK

As it often happens in human gross anatomy, the names of the vessels and nerves of the root of the neck will help students to understand and memorize the structures they need to know. To begin with, it makes sense that the *vertebral artery* is a deep artery that passes inside the transverse foramina of cervical vertebrae, and that the *internal thoracic artery* passes internally to the rib cage. Regarding the other (3rd) branch of the 1st part of the subclavian artery, its name *"thyrocervical trunk"* announces that it has *thyroid* (the inferior thyroid artery) and *cervical* (the transverse cervical artery) components. Moreover, the name *"inferior thyroid artery"* announces that there is a **superior thyroid artery**; and that because this latter artery does not branch from the subclavian artery, it has to be derived from the **common carotid artery**, as will be seen in the next paragraphs. The names of the third branch of this thyrocervical trunk and the single branch of the third part of the subclavian artery are more difficult to remember, but an easy way to memorizing them—if you have already studied the upper limb—is to remember that their order is *opposite* to that of the respective nerves derived from the **brachial plexus**. That is, in the brachial plexus, the **dorsal scapular nerve** branches before (or medially to) the branching of the **suprascapular nerve**, whereas in the arteries, the suprascapular artery branches before (medially to) the dorsal scapular artery.

**BOX 3.18   EASY WAYS TO REMEMBER: SIX BRANCHES
OF THE EXTERNAL CAROTID ARTERY**

These six branches are easy to learn using the simple schematic drawing
shown in Figure 3.5. The three major external prominent features of the
anterior region of the head and neck are the **face**, the **tongue**, and the
thyroid gland, whereas the three major features more posteriorly are

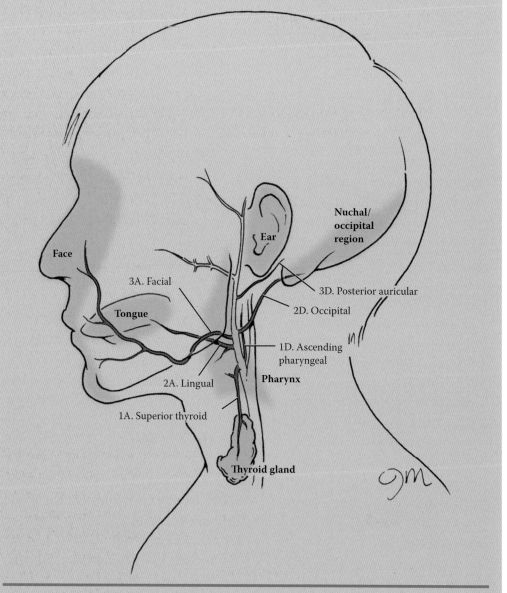

Figure 3.5   Branches of the external carotid artery.

the **ear**, the nuchal/**occipital region**, and the **pharynx**. Accordingly, the three anterior branches of the external carotid artery are, from inferior to superior, the **superior thyroid artery** (logically, because we already know that the inferior thyroid artery branches from the subclavian artery), the **lingual artery**, and the **facial artery**. Remember that a major difference between the facial and lingual arteries is that the facial is very superficial, whereas the lingual passes *deep* to the major muscle landmark of the superior neck, the **hyoglossus muscle** (see Box 3.10). The three posterior branches of the external carotid artery, from inferior to superior, are the **ascending pharyngeal artery** ("ascending" makes sense because it branches so inferiorly, that has to ascend to reach the major portion of the pharynx), the **occipital artery**, and the **posterior auricular artery**. Moreover, the anterior and posterior branches usually alternate, so the ascending pharyngeal branches between the superior thyroid and the lingual, then the occipital branches between the lingual and facial, and lastly the posterior auricular branches superior to the facial, before the external carotid bifurcates superiorly into the maxillary and superficial temporal arteries.

At the level of the mouth, the facial artery gives rise to the **inferior labial artery** and the **superior labial artery**, and then becomes the **angular artery** at the level of the nose. The superior thyroid artery gives rise to the **superior laryngeal artery**, which pierces the **thyrohyoid membrane** together with the **superior laryngeal vein** and the **internal laryngeal nerve** to enter the **larynx**. Some students get confused about why the nerve and artery piercing the thyrohyoid membrane have different names, such as internal laryngeal nerve and superior laryngeal artery. However, keep in mind that the internal laryngeal nerve comes from the **superior laryngeal nerve**, so that the name is similar to that of the artery; the difference being that the superior laryngeal nerve also gives rise to the **external laryngeal nerve**, which innervates the inferior pharyngeal constrictor and cricothyroid muscles.

The **carotid sinus** is a dilation of the origin of the **internal carotid artery** at the level of the bifurcation of the common carotid artery. The wall of the carotid sinus, which is innervated by the **glossopharyngeal nerve** (**CN IX**), contains baroreceptors that monitor blood pressure. If necessary (such as in an attack of tachycardia; i.e., rapid heartbeat) a doctor may apply strong pressure to this carotid receptor to quickly slow down a patient's heart rate (however, this intervention is risky because blood pressure and heart rate may decrease too quickly and the patient may faint; so please don't try this at home). The carotid body is a small mass of tissue located at the medial aspect of the carotid bifurcation, also innervated by the glossopharyngeal nerve (CN IX), which acts as a chemoreceptor by monitoring changes in oxygen and carbon dioxide concentration of the blood. The **lingual artery** passes deep to the hyoglossus muscle and then gives rise to the **deep lingual artery** that supplies part of the tongue (Plate 3.41).

The **maxillary artery** (Plate 3.29) is superficial to the **lateral pterygoid muscle** in about two-thirds of human adults and deep to this muscle in about one-third of human adults. One of its major branches is the **middle meningeal artery**, which arises medial to the neck of the mandible, then runs superiorly to pass between fibers of the **auriculotemporal nerve** before entering the **foramen spinosum** to supply the **dura mater** (Plate 3.22). Anteriorly to the middle meningeal artery, and also running superiorly, are the **anterior and posterior deep temporal arteries** that enter the deep surface of the **temporalis muscle**, and the **masseteric artery** that enters the deep surface of the **masseter muscle**. In contrast, the **inferior alveolar artery** runs inferiorly to enter the **mandibular foramen** with the **inferior alveolar nerve**. The **buccal artery** runs anteriorly and inferiorly to supply the **cheek**. After giving rise to these branches, the maxillary artery runs anteriorly toward the **pterygopalatine fossa** and divides into four terminal branches: the **posterior superior alveolar artery**, the **infraorbital artery, descending palatine artery**, and **sphenopalatine artery**. The **sphenopalatine artery** runs with the **nasopalatine nerve** from the sphenopalatine foramen to the **incisive canal** and supplies the nasal septum and part of the mucosa that covers the anterior hard palate. The sphenopalatine artery branches into the **posterior lateral nasal artery** that goes to the lateral nasal wall and the **posterior septal branch** that goes to the nasal septum. The **descending palatine artery** lies in the **greater palatine canal** that runs vertically, lateral to the palatine bone, from the region of the sphenopalatine foramen to the greater palatine foramen. Distally, the descending palatine artery divides into the **greater palatine artery** and the **lesser palatine artery** to supply the hard and soft palates respectively. The other terminal branch of the maxillary artery that lies in this region, the **infraorbital artery**, passes across the inferior orbital fissure, enters the infraorbital canal, and then emerges on the face through the infraorbital foramen.

The **ophthalmic artery** branches from the **internal carotid artery** and passes into the orbit with the optic nerve (CN II). It forms the **central artery of the retina**, then divides into the lacrimal, the supratrochlear, and supraorbital arteries, the latter two of which appear on the external surface of the orbital region (Plates 3.22 and 3.34).

3.3.2.2 Head and Neck Veins

BOX 3.19   BASIC CONCEPTS IN VENOUS DEVELOPMENT

In vertebrate embryos, including human embryos, most veins are initially bilaterally symmetrical (Plate 3.6). This symmetry is retained in most head, neck, body wall, and limb veins but is lost during development of thoracic and abdominal veins (Plate 3.6). In general, embryonic veins on the *right* side are more likely to be retained than those on the left. A series of **cardinal veins** form early and drain all body tissues except those associated with the gut, which are drained by **vitelline veins** that carry blood from the yolk sac and intestines to the liver. Adult veins of the head (branches of the **jugular veins**) and trunk (branches of the **inferior vena cava**) are derived from these cardinal veins. During embryonic/fetal development, blood carrying oxygen and nutrients from the placenta flows via the **left umbilical vein** directly to the **liver**, then through an enlarged intrahepatic channel, the **ductus venosus**, directly to the left hepatic vein and into the inferior vena cava. The ductus venosus allows enriched blood to bypass hepatic sinusoids and go directly to the inferior vena cava and **heart**. The ductus venosus collapses when this placental flow ceases (i.e., after birth).

In the embryo, paired, multibranched **cranial cardinal veins** drain blood from the head (Plate 3.6). These cranial cardinal veins join with caudal cardinal veins from the trunk and, together, form bilateral **common cardinal veins** that enter the paired horns of the **sinus venosus**. An anastomosis develops between the left and right cranial cardinal veins. At the same time, venous drainage from both head and trunk somatic tissues shifts toward the right side. Subsequently, the left cranial cardinal degenerates between the anastomosis and the left common cardinal vein, and the anastomotic branch becomes the **left brachiocephalic vein**. The segment of original right cranial cardinal between the junction of the two brachiocephalics and the heart (right atrium) is the **superior vena**

cava. As the heart wall expands, the left common cardinal vein becomes incorporated into its wall as the **coronary sinus**, the site of drainage for most **coronary veins**.

In the trunk, the embryo forms three sets of veins. The aforementioned paired **cardinal veins**, which drain all somatic tissues plus kidneys and many urogenital structures, are the first and most complex of these. New and more complex kidneys arose in tetrapods (animals with upper and lower limbs) with the switch to terrestrial living and new veins were needed. Tetrapod embryos evolved a new pair of veins that run the entire length of the trunk, connecting with preexisting cardinal veins near their cranial and caudal ends, and also forming many local anastomoses with intersegmental branches of caudal cardinal veins. Whenever new and old venous drainages became redundant, the older vessels degenerated. This rearrangement of veins creates new channels that include parts of both older and newer veins. Veins tend to have many more anastomoses and interconnections than arteries; remodeling of these tortuous channels during development also creates more variations of adult veins than is present in arteries. The same remodeling process also occurred during the evolution of terrestrial vertebrates. As a result, the **caudal vena cava**, which is the principal vein draining abdominal somatic tissues, is derived from three sets of cardinal veins and their anastomoses.

The second set of abdominal veins is the **vitelline veins**, which drain the stomach, intestines, and yolk sac. These veins do not drain directly into the cardinal veins, but rather flow directly into the liver as the **hepatic portal veins**, where they are broken up into hepatic sinusoids. After filtration, blood leaves the liver via hepatic veins that drain into the caudal vena cava.

The third venous system is the **umbilical vein**, which brings placental blood to the heart. It passes through the liver, but is not broken up into microscopic sinusoids.

As noted in Box 3.19, there are some similarities between the veins and arteries in the adult human head and neck, but in general they are more different from each other than are the veins and arteries in the limbs. Therefore, as we do for the arteries of the head, we will treat the veins branches of the **superior vena cava** to the different regions of the head and neck, even though the blood obviously runs the other way, draining blood from head to heart. We begin with the two major branches of the superior vena cava: (1) the **right brachiocephalic vein** that gives rise to the **right subclavian vein** and the **right internal jugular vein** and (2) the **left brachiocephalic vein** that gives rise to the **left subclavian vein** and the **left internal jugular vein**. This branching pattern is the very first major difference between veins and arteries: As noted in Section 3.3.2.1, there is no left brachiocephalic artery, another example of body asymmetry.

Another major difference between head/neck arteries and veins is that, although the **internal jugular vein** gives rise to three major anterior branches, as does the external carotid artery, its branches differ from those of this latter artery. The branches of the internal jugular vein are the **middle thyroid vein**, the **superior thyroid vein** which is associated with the **superior laryngeal vein**, and the **common facial vein** which gives rise to the **facial vein** and the **lingual vein**. In contrast, the three anterior branches of the external carotid artery are the superior thyroid artery (which gives rise to the superior laryngeal artery), lingual artery, and facial artery. You can remember the differences between the anterior branches of the internal jugular vein and of the external carotid artery this way (compare Plates 3.23 and 3.21):

> The internal jugular vein gives rise to a vein that has no equivalent in the arterial system: the middle thyroid vein (there is no middle thyroid artery); to compensate for this extra vein, there is an extra artery coming from the external carotid artery, while the internal jugular veins gives rise to a single common facial vein that then bifurcates into facial and lingual veins, the external carotid artery directly gives rise to a facial artery and to a lingual artery.

Another major difference between head and neck veins and arteries is that the posterior auricular artery is a posterior branch of the external carotid artery, whereas the **posterior auricular vein** is one of the two main superior branches of the **external jugular vein**, together with the **retromandibular vein** (this latter vein joins branches of the **common facial vein**). No arteries clearly correspond to the external or anterior jugular veins, which are essentially vertical, superficial veins situated, respectively, on the lateral and anterior surfaces of the neck, and which usually branch from the subclavian vein. In yet another difference between arteries and veins, the lingual vein is superficial to the **hyoglossus muscle** and the subclavian vein is superficial to the **anterior scalene muscle**, whereas the lingual artery is deep to the hyoglossus muscle and the subclavian artery is deep to the anterior scalene muscle. However, these two differences can be explained by the rule that, in general, veins are superficial to arteries in the head and neck just as they are in the limbs.

Of course, there are also many similarities between the veins and arteries: There is an inferior thyroid artery branching from the subclavian artery and an **inferior thyroid vein** usually branching from the **brachiocephalic vein**; the occipital artery is a posterior branch of the external carotid artery and the **occipital vein** is a posterior branch of the internal jugular vein; the facial vein gives rise to the **inferior labial vein, superior labial vein**, and **angular vein**, and the two major superior branches of the superficial venous system of the head are the **maxillary veins** and the **superficial temporal vein**, in a configuration very similar to that seen in the arterial system. Note however that the **superior ophthalmic vein** and the **inferior ophthalmic vein** run respectively superior and inferior to the eye, the superior ophthalmic vein anastomosing with the **angular vein**, which is a branch of the facial vein (Plate 3.22). This anastomosis is clinically important because infections of the nasal cavity, cheeks, forehead, and upper lip can be spread via the facial vein through the angular and superior ophthalmic veins to the cavernous sinus, a dural venous sinus in the base of the cranium, and result in thrombosis of this sinus. This condition affects the abducens nerve and, subsequently, the lateral rectus muscle and the movements of the eye.

3.4 Head and Neck Muscular System

3.4.1 Muscles of 1st Arch (Nerve CN V)

All the eight muscles of the 1st arch are innervated by the mandibular division (CN V$_3$) of the trigeminal nerve (CN V) (Plate 3.2). The evolutionary and developmental subgroups of the musculature of this 1st arch are listed in Table 3.1. Both the **anterior digastric** and the **mylohyoid** derive developmentally from the inferior (ventral) portion of the mandibular (1st arch) muscle primordium, so it makes sense that both these muscles are innervated by the mylohyoid nerve. In contrast, the **masseter, temporalis, lateral pterygoid muscle**, and **medial pterygoid muscle** derive from the main portion of the mandibular (1st arch) muscle primordium, as do the **tensor tympani** and **tensor veli palatini,** so all these muscles are more closely related to each other evolutionarily and developmentally than they are to the mylohyoid muscle and to the anterior digastric muscle. The medial pterygoid, although it is functionally more closely related to the lateral pterygoid, masseter and temporalis in adult humans, seems to be developmentally and evolutionarily

Table 3.1 **Evolutionary and Developmental Subgroups of the Musculature of the 1st Arch in Humans**

Evolutionary/Developmental Subgroups	Muscles of 1st (Mandibular) Arch
Ventral muscles	Mylohyoid
	Anterior digastric
More lateral subgroup derived from "adductor mandibulae" complex of fishes (namely from adductor mandibulae A2)	Masseter
	Temporalis
	Lateral pterygoid
More medial subgroup derived from "adductor mandibulae" complex of fishes (namely from adductor mandibulae A3)	Medial pterygoid
	Tensor tympani
	Tensor veli palatini

more closely related to the tensor tympani and tensor veli palatini. The developmental/evolutionary origins of some of the muscles in this section, shown in Table 3.1, contrast with the functional groupings often given in textbooks. We use the functional groupings in the text to maintain consistency with other anatomical texts, and because some of them are clinically useful.

3.4.1.1 Muscles Directly Involved with Movements of Mandible

The **anterior digastric muscle** attaches anteriorly onto the **digastric fossa of the mandible** and is innervated by the **mylohyoid nerve** (Plates 3.25a and 3.29). This muscle is attached posteriorly by the **intermediate digastric tendon** to the posterior digastric muscle, which is a 2nd arch muscle innervated by the facial (CN VII) nerve, so the entire muscle complex is called the "digastric," or two-bellied muscle. The digastric muscle complex elevates the **hyoid bone** and depresses the **mandible** (opens the jaw), the posterior digastric being mainly related to the former movement, and the anterior digastric to the latter movement. This is logical, because the posterior digastric is derived from the 2nd arch, and thus originally associated with movements of the hyoid apparatus; in humans it maintains a connection with the hyoid bone via a fibrous sling that holds the intermediate digastric tendon to the body of the hyoid. The **mylohyoid muscle** is also innervated by the mylohyoid nerve and mainly connects the two sides of the mandible—reflecting its developmental derivation from the 1st arch—although it also attaches on to the hyoid bone. In humans the demi-mandibles are fused. Therefore, the mylohyoid has lost its main original function, which was to connect and bring together the two demi-mandibles (as illustrated by the name of the muscle complex that gave rise to this muscle in evolution—the intermandibularis muscle complex). In our species, the mylohyoid is mainly involved in depression of the mandible, raising the floor of the mouth during the first stage of swallowing, and elevation of the hyoid bone.

The **masseter muscle** is innervated by the **masseteric nerve** (Plate 3.15), that runs from the **zygomatic arch** to the lateral surface of the **mandibular ramus** and elevates the mandible (closes the jaw) and protrudes the mandible (moves the jaw forward). The other main muscle of mastication, the temporalis, is innervated by the **deep temporal nerves** and runs from the **temporal fossa** (it is attached to the deep surface of the **temporal fascia**) to the **coronoid process of the mandible**. The anterior fibers of the temporalis have a more vertical orientation and mainly elevate the mandible, whereas its posterior fibers have a more horizontal orientation and mainly retrude the mandible (move the jaw backward). The **lateral pterygoid muscle** is deep to the mandible. The superior head of the lateral pterygoid muscle originates from the infratemporal surface of the greater **wing of the sphenoid bone**, and the inferior head originates from the lateral surface of the **lateral plate of the pterygoid process** of the sphenoid bone. The lateral pterygoid inserts onto the **mandibular neck** (mainly the inferior head of the muscle) and **articular disc** within the capsule of the **temporomandibular joint** (mainly the superior head of the muscle), and its function is to depress the mandible. The **medial pterygoid muscle** lies deep to the lateral pterygoid, running from the maxilla and medial surface of the lateral plate of the pterygoid process to the inner surface of the ramus of the mandible. Its function is to elevate the mandible. Of the four major muscles of the temporal region (masseter, temporalis, medial, and lateral pterygoid muscles), *only the lateral pterygoid helps open the mouth*. The insertion of the lateral pterygoid on the mandibular neck pulls the neck of the mandible forward, out of the mandibular fossa and on to the articular eminence. Only then is the mandible free to rotate in its lower synovial cavity, without butting against posterior structures (Plate 3.28). Unless there is resistance, the downward rotation of the jaw is accomplished mostly by gravity.

3.4.1.2 Tensor Tympani and Tensor Veli Palatini

The **tensor veli palatini** is a 1st arch (mandibular) muscle, accordingly innervated by the trigeminal nerve (CN V), but actually lies in the pharyngeal region (Plate 3.36). True to its developmental origin as a muscle attached to part of the cartilage of the 1st arch, it is more superficial (lateral) than true pharyngeal muscles such as the levator veli palatini. In the adult, the tensor veli palatini originates from the scaphoid fossa of the sphenoid bone and wraps around the hamulus of the medial pterygoid plate to attach onto the palatine aponeurosis, and therefore tenses the soft palate when contracted. The **tensor tympani**, a middle ear muscle attached to the handle of the malleus (also a 1st arch derivative), is also innervated by the trigeminal nerve (CN V) (Plate 3.45). As its name indicates, this muscle tenses the tympanic membrane to dampen loud sounds. *Therefore, keep in mind that the two head muscles that have names including "tensor" are both 1st arch (mandibular) muscles, innervated by CN V.*

3.4.2 Muscles of 2nd Arch (Nerve CN VII)

All the muscles of the 2nd arch are innervated by the facial nerve (CN VII). The evolutionary and developmental subgroups of the musculature of 2nd arch are shown in Table 3.2. As can be seen in this table, the functional groupings listed in the subsections below (for instance facial expression versus nonfacial expression muscles) do not seem to correspond to true developmental/evolutionary groups.

3.4.2.1 Muscles of Facial Expression

The muscles of facial expression are only present in mammals, and this is one of the few muscle groups where humans have more muscles than most other mammals. The evolutionary increase in number of facial muscles in humans compared to even our closest living relatives—the great apes— reflects the acutely important role played by nuanced facial expression and thus verbal and nonverbal communication in the evolution of our highly social species. Evolutionarily and developmentally, human facial expression muscles are derived from muscle progenitors that were originally associated with skeletal structures of the 2nd (hyoid) arch, but which secondarily migrated to other regions of the head and neck, and even to parts of the pectoral region (the platysma). Like all other 2nd arch muscles, the muscles of facial expression receive motor innervation from the facial nerve (CN VII).

The muscles of facial expression are shown in Plates 3.27 and 3.16. The **platysma muscle** runs from the upper pectoral region to the inferior border of the mandible, skin of the cheek, and angle of the mouth. The **orbicularis oculi** encircles the **palpebral fissure** (opening of the eyelid) and has

Table 3.2   Evolutionary and Developmental Subgroups of the Musculature of the 2nd Arch in Humans

Evolutionary/Developmental Subgroups	Muscles of 2nd (Hyoid) Arch
Muscles derived from the "depressor mandibulae" group	Stylohyoid + posterior digastric
	Platysma + risorius
	Occipitalis + posterior auricular + part of external auricular muscles
Muscles derived from the "levator hyoideus" group	Stapedius
Muscles derived from the "sphincter colli profundus/cervicalis transversus" group	Zygomaticus major
	Zygomaticus minor
	Frontalis + anterior auricular + superior auricular + temporoparietalis
Muscles derived from the "orbicularis oculi" group	Orbicularis oculi
	Depressor supercilii + corrugator supercilii
Muscles derived from the "naso-labialis" group	Levator labii superioris alaeque nasi + procerus
Muscles derived from the "buccinatorius" group	Buccinatorius
Muscles derived from the "orbicularis oris" group	Levator labii superioris
	Nasalis
	Depressor septi nasi
	Levator anguli oris
	Orbicularis oris + depressor labii inferioris + depressor anguli oris
Muscles derived from the "mentalis" group	Mentalis

three parts: The **orbital part** surrounds the orbital margin and is involved in tight closure of the eyelid; the **palpebral part** is contained in the eyelids and is involved in the blinking of the eyelid; and the **lacrimal part** that lies deep to the palpebral part, pressing on the lacrimal sac. The **levator labii superioris** runs from the maxilla to the upper lip and elevates the upper lip. This lip muscle is just lateral to the **levator labii superioris alaeque nasi** that runs from the upper lip to the superior region of the nose (just lateral to the inferior portion of the **procerus**). The **zygomaticus major** runs from the zygomatic bone to the angle of the mouth and, together with the **zygomaticus minor**, draws the angle of the mouth superiorly and posteriorly (i.e., in a smile). The **orbicularis oris** is attached to the maxilla, mandible, skin of the mouth, and angle of the mouth, and functions as the sphincter of the mouth. The **depressor anguli oris** lies superficial to the **depressor labii inferioris** (which in turn lies mainly superficial to the **mentalis**) and runs from the mandible to the angle of the mouth to depress the corner of the mouth (i.e., in a frown).

The **occipitofrontalis** muscle is formed by the frontalis and the occipitalis muscles, connected by the **epicranial aponeurosis** that lies between the superficial layers (**skin** and **connective tissue** formed by dense subcutaneous tissue containing the vessels and nerves of the scalp) and deep layers (**loose connective tissue** that permits the scalp to move over the calvaria and **pericranium**, that is the periosteum of the cranial bones) of the **scalp**. Near to the **occipitofrontalis** lie the four facial muscles connecting the skull to the external ear, which in humans (contrary to most mammals) usually only produce minor movements of the ear lobe. These four muscles are the **auricularis anterior**, **auricularis posterior**, **auricularis superior**, and the **temporoparietalis** (a thin muscle usually not described or shown in atlases of human anatomy). Other facial muscles present in humans and most mammals are the **depressor supercilii** and **corrugator supercilii** above the eye; the **nasalis** and **depressor septi nasi** in the nose region; and the deep muscles **levator anguli oris** and **buccinator**. The buccinator runs from the **pterygomandibular raphe** and the lateral surfaces of the alveolar processes of the maxilla and mandible to the angle of the mouth. The buccinator is a peculiar muscle of facial expression because it is the only facial muscle that also significantly helps with **mastication** by compressing the cheek against the molar teeth, thus keeping food on the grinding surfaces of the molar teeth during **chewing**. In addition, humans have a muscle that is almost unique to our species: the **risorius**, attached to the corner of the mouth (it is sometimes also present in great apes such as chimpanzees). Its name comes from the Latin "risus" meaning "laugh," as this muscle pulls the corners of the lips laterally and posteriorly.

3.4.2.2 Stylohyoid, Posterior Digastric, and Stapedius

The **posterior digastric muscle** attaches posteriorly onto the **mastoid process of the temporal bone** and anteriorly by the **intermediate**

digastric tendon to the **anterior digastric muscle** (Plates 3.24, 3.25a, and 3.35), which as noted above is a 1st arch muscle structure innervated by the trigeminal nerve. The intermediate tendon is attached to the **body of the hyoid bone** and to the **greater horn of the hyoid bone** by a **fibrous sling**. This sling reveals the original attachment of the muscle structure that gave rise to the posterior digastric muscle: the hyoid bone, which is logically the main embryonic attachment of 2nd arch (hyoid) muscles. As a whole, the complex formed by the anterior and posterior digastric muscles elevates the **hyoid bone** and depresses the **mandible**, as explained above. As its name indicates, the **stylohyoid muscle** runs from the **styloid process of the temporal bone** to the body of the hyoid bone, thus keeping its embryonic attachment to the hyoid bone. Thus, the stylohyoid muscle logically functions to raise the hyoid bone.

Although the **stapedius** is a middle ear muscle in mammals, including humans, it keeps its original attachment to structures of the 2nd (hyoid) arch because, as its name indicates, the stapedius attaches onto the neck of the **stapes**, a bone derived from the 2nd arch (the stapedius originates from the **pyramidal eminence** of the middle ear cavity) (Plate 3.45a and c). The stapedius is innervated by a branch of the facial nerve, the **nerve to stapedius**, and controls the amplitude of sound waves to the **inner ear**.

3.4.3 Muscles of More Caudal Branchial Arches (Nerves CN IX, CN X, and CN XI)

All the muscles of the more caudal branchial arches are innervated by the glossopharyngeal (CN IX), vagus (CN X), and accessory (CN XI) nerves (Plate 3.35). The evolutionary and developmental subgroups of the musculature of the more caudal arches are shown in Table 3.3. As can be seen in this table, the functional subgroupings listed in the three subsections below mainly correspond to developmental/evolutionary subgroupings, with the exception of the grouping of the stylopharyngeus with the pharyngeal muscles. However, functional groups vary among textbooks; for example, some group the cricothyroid muscle with the laryngeal and not with the pharyngeal group, which is its true developmental/evolutionary group. Mapping studies in bird embryos performed by Noden and colleagues have suggested that the true laryngeal muscles may be derived from somites, whereas other studies in mice strongly suggest a nonsomitic—and thus a branchiomeric—origin. More studies are needed to clarify these seemingly contradictory data (see Plate 3.2), but for the time being, we will use the terminology that is

Table 3.3 Evolutionary and Developmental Subgroups of the Musculature of the More Caudal Branchial Arches in Humans

Evolutionary/Developmental Subgroups	Muscles of the More Caudal Branchial Arches
"True branchial muscles" (Note: All the "true branchial muscles *sensu stricto*" were lost in the evolution of fishes and tetrapods, with exception of the stylopharyngeus; the trapezius and sternocleidomastoid are "true branchial muscles" but not *sensu stricto*, that is they were instead derived from the cucullaris group)	Stylopharyngeus
	Trapezius + sternocleidomastoid
Pharyngeal muscles (Note: The palatopharyngeus and musculus uvulae are closely related to each other)	Superior, middle and inferior pharyngeal constrictors and cricothyroid
	Levator veli palatini + salpingopharyngeus + patalopharyngeus and musculus uvulae
Laryngeal muscles (Note: The vocalis derives from the thyroarytenoideus, and the transverse arytenoid and oblique arytenoid are closely related to each other)	Cricoarytenoideus posterior
	Thyroarytenoideus and vocalis + cricoarytenoideus lateralis + arytenoideus transversus and obliquus

more commonly used for mammals and in particular for humans and thus refer to the "true laryngeal muscles" as branchiomeric muscles (Figure 3.1).

3.4.3.1 Stylopharyngeus (Nerve CN IX) and True Pharyngeal Muscles (Nerve CN X)

The **pharyngeal wall** has three layers, which are, from superficial to deep: (1) the **buccopharyngeal fascia**, which is continuous with the connective tissue that covers the buccinator muscle; (2) the **muscular layer** composed of an outer circular part (**superior, middle, and inferior pharyngeal constrictors**) and an inner longitudinal part (**salpingopharyngeus** and **palatopharyngeus**, fused inferiorly with the **stylopharyngeus**); and (3) the **pharyngobasilar fascia**, which as its name indicates connects the superior constrictor muscle to the base of the skull. All the muscles of the muscular layer are innervated by the **vagus nerve (CN X)**, except the stylopharyngeus, which is the only skeletal muscle of the body that receives motor innervation from the **glossopharyngeal nerve (CN IX)**, as explained above. The **inferior pharyngeal constrictor** runs from the **pharyngeal raphe** to the thyroid cartilage (**thyropharyngeal muscle bundle**) and the cricoid cartilage (**cricopharyngeal muscle bundle**) (Plate 3.35). The superior part of the inferior constrictor lies superficial to the **middle pharyngeal constrictor**, which runs from the pharyngeal raphe to the greater horn of the hyoid bone and the inferior portion of the stylohyoid ligament (Plate 3.36b). The superior part of the middle constrictor lies superficial to the **superior pharyngeal constrictor**, which runs from the **pharyngeal tubercle of the occipital bone** and the pharyngeal raphe to the pterygomandibular raphe (**buccopharyngeal muscle bundle**), the **hard palate** (**pterygopharyngeal muscle bundle**), the **tongue** (**glossopharyngeal muscle bundle**) and the **mandible** (**mylopharyngeal muscle bundle**) (Plate 3.36c). The **stylopharyngeus** muscle and the glossopharyngeal nerve (CN IX) that innervates it enter the pharynx by passing between the middle and superior constrictors.

The **musculus uvulae** lies in the midline of the soft palate, anterior to the **palatopharyngeus**. The palatopharyngeus runs from the hard palate and palatine aponeurosis to the thyroid cartilage and pharyngeal wall, elevates the larynx during swallowing, and forms part of the **longitudinal constrictor layer of the pharynx** together with the salpingopharyngeus and the stylopharyngeus. The **salpingopharyngeus** originates from the cartilage of the auditory tube. Then, the salpingopharyngeus blends with the palatopharyngeus to attach inferiorly onto the thyroid cartilage and pharyngeal wall, thus also helping to raise the larynx during swallowing. The **levator veli palatini** runs from the cartilage of the auditory tube and adjacent temporal bone to the palatine aponeurosis of the soft palate, thus lifting the soft palate. The **palatoglossus** runs from the palatine aponeurosis to the lateral side of the tongue, and thus elevates the tongue and depresses the soft palate, and is also innervated by the vagus nerve (CN X). As such, contrary to what its name indicates, the palatoglossus is a pharyngeal muscle, not a true tongue muscle (and thus not a hypobranchial, somitic muscle).

3.4.3.2 *True Laryngeal Muscles (Recurrent Laryngeal Nerves)*

As noted above, based largely on studies of other animals such as chickens, some authors have suggested that the laryngeal muscles might not be true head (branchiomeric) muscles but instead might derive from somites as do the infrahyoid and tongue muscles (see Section 3.4.4). However, both classical embryological studies such as Edgeworth's still outstanding 1935 book, and more recent studies using contemporary genetic and genomic tools have strongly supported the idea that the laryngeal muscles and even at least part of the **esophageal muscles** are in fact branchiomeric muscles derived from the most caudal branchial arches (Plate 3.2).

The **larynx** is involved in both producing sound during phonation (mainly controlled by the **true laryngeal muscles** (Plates 3.42 through 3.44), which are often designated "**intrinsic laryngeal muscles**") and controlling the airway. The position of the larynx is actually controlled mainly by other, nonlaryngeal muscles that are often designed "**extrinsic laryngeal muscles**" in the literature. A common mistake that students make is to consider the **cricothyroid muscle**—which connects the cricoid and thyroid cartilages and lengthens the vocal fold—a true laryngeal muscle. In fact, the cricothyroid muscle is developmentally and evolutionary part of the pharyngeal musculature, being derived from the inferior pharyngeal constrictor muscle and innervated, like the latter muscle, by the **external laryngeal nerve**. The true laryngeal muscles are all *inside* the larynx and are innervated accordingly by the **inferior laryngeal nerve**. This nerve, situated below the glottis and above the cricothyroid joint, is a communicating branch between the **internal laryngeal nerve** (which innervates the laryngeal mucosa above the glottis), and the **recurrent laryngeal nerve** (situated below the cricothyroid joint).

Therefore, there are five true laryngeal muscles, and their names are easy to remember because they all refer to the respective attachments of each muscle. First, the **posterior cricoarytenoid** muscle runs superolaterally from the posterior surface of the cricoid cartilage to the lateral, muscular process of the arytenoid cartilage. It is the only abductor of the vocal folds; developmentally and evolutionary, it is not included in the subgroup that includes the other four true laryngeal muscles. Second, the **lateral cricoarytenoid** muscle runs superomedially from the anterolateral surface of the cricoid cartilage to the muscular process of the arytenoid cartilage, and therefore contributes to the adduction of the vocal folds. Third, the **arytenoid muscle complex** includes a **transverse arytenoid muscle** with transverse fibers, and a **oblique arytenoid muscle** with oblique fibers; these two muscles mainly connect the arytenoid cartilages across the midline and therefore also adduct the vocal folds. However, the oblique fibers can extend superiorly to attach onto the epiglottis via **aryepiglottic muscle fibers** that are sometimes present in humans. Fourth, the **thyroarytenoid** muscle connects the inner portion of the anterior aspect of the thyroid cartilage to the arytenoid cartilage and thus moves this latter cartilage anteriorly, relaxing the vocal folds, and opposing the action of the cricothyroid muscle. Fifth, the very thin **vocalis muscle** is often said to be present only in humans and associated to our ability to speak; however, in reality the vocalis muscle is also found in various other mammals. The vocalis corresponds to the medial fibers of the thyroarytenoid muscle that do not attach onto the thyroid and arytenoid cartilages, but attach mainly onto the vocal ligament and thus modify the tension of this ligament, modulating pitch.

3.4.3.3 Neck Muscles Trapezius and Sternocleidomastoid (Nerve CN XI)

The **trapezius** is usually described as a part of the "**superficial muscles of the back**" (Plate 4.6) but evolutionarily and developmentally it is a head muscle. The trapezius and the **sternocleidomastoid** receive motor innervation via a cranial nerve (**accessory nerve**; **CN XI**), and originate embryologically from the region of the posterior branchial arches, subsequently expanding caudally to attach onto the **pectoral (or shoulder) girdle (scapula** and **clavicle**) and **sternum**. During their caudal expansion, they form connections with the cervical spinal nerves C3 and C4, which provide sensory innervation to them. This pattern of innervation, together with their adult attachments onto the shoulder girdle and sternum, suggest that the trapezius and sternocleidomastoid should be considered head muscles that share features with postcranial muscles. These muscles help to blur the distinction between head and trunk (see also Box 3.2). In fact, the trapezius and sternocleidomastoid played an important role in the origin and early evolution of the **neck**, which separated the head from the region of the upper limbs in tetrapods, that is, animals with upper and lower limbs (amphibians, reptiles, and mammals) (see Box 3.14).

The human trapezius has superior, middle, and inferior heads, which can be contracted separately. This separation reflects the fact that, in our ancestors, the trapezius complex actually included three distinct muscles, which can be still seen in many modern mammals. During human evolution, these three muscles fused into a single trapezius muscle. The superior head runs from the occipital bone of the skull to the lateral third of the clavicle and might reach the lateral portion of the scapula. Therefore, apart from elevating the scapula, the superior head can laterally rotate the scapula. This rotation, as well as its indirect contribution to the abduction of the arm are described in Section 4.3.1. The middle head of the trapezius runs mainly from the nuchal ligament to the back of the acromion and spine of the scapula and thus retracts the scapula. The inferior head originates mainly from the spinous processes of C7 to T12 and inserts near the medial end of the scapular spine, depressing the scapula.

The sternocleidomastoid runs from the mastoid process of the temporal bone and the superior nuchal line to the manubrium of the sternum (sternomastoid head) and to the medial portion of the clavicle (cleidomastoid head) (Figure 3.6, Plate 3.25). When the sternocleidomastoid muscle of one side of the body acts alone, it rotates the head to the opposite side (contralaterally) and flexes the neck to the same side (ispilaterally). When the sternocleidomastoid muscles of the two sides of the body act together, they can flex the neck or extend the head (because their fibers pass anterior to most of the joints between cervical vertebrae but posterior to the atlanto-occipital joint).

3.4.4 Hypobranchial Muscles of Tongue and Infrahyoid Region (Cervical Nerves and CN XII)

The tongue has both intrinsic and extrinsic muscles. Intrinsic muscles are found completely within the body of the tongue (Plate 3.40) and do not have attachments to skeletal structures. Extrinsic muscles course between the tongue and surrounding skeletal structures. Both sets are critical for moving food to the pharynx, swallowing, and in humans for speech. The muscles of the tongue and infrahyoid region are called **hypobranchial muscles** and are very different from the muscles described in Sections 3.4.1, 3.4.2, 3.4.3, and 3.4.5 because they are not true head muscles. They instead have a **somitic origin**, that is, they are trunk muscles that during human evolution and development have migrated rostrally to the neck and head. Their somitic origin explains why all four true extrinsic tongue muscles and all four infrahyoid muscles are innervated by cervical nerves or by the **hypoglossal nerve**, which is considered to be the last cranial nerve (**CN XII**) but is deeply associated with the **spinal nerves** (Plate 3.19), as explained in Section 3.3.1.12. The evolutionary and developmental subgroups of the hypobranchial musculature are shown in Table 3.4. However, it should be noted that recent studies have suggested that part of the extrinsic tongue musculature might have a branchiomeric (not somitic) origin (see Plate 3.2).

3.4.4.1 Infrahyoid Muscles (Cervical Nerves)

Three of the infrahyoid muscles, the **sternohyoid**, **sternothyroid**, and **omohyoid** are innervated by branches of the **ansa cervicalis**, whereas the

Table 3.4 **Evolutionary and Developmental Subgroups of the Hypobranchial Musculature in Humans**

Evolutionary/Developmental Subgroups	Hypobranchial Muscles
Tongue muscles (Note: The styloglossus and hyoglossus are closely related to each other)	Geniohyoid
	Genioglossus + part of intrinsic tongue muscles
	Hyoglossus + part of intrinsic tongue muscles + styloglossus
Infrahyoid muscles	Omohyoid
	Sternohyoid + sternothyroid + thyrohyoid

most superior infrahyoid muscle, the **thyrohyoid**, is innervated by **nerve C1**. A branch of this nerve runs or "hitchhikes" with the hypoglossal nerve until it gives rise to the **nerve of the thyrohyoid muscle**. In accord with the strong evolutionary and developmental relationship between the tongue and infrahyoid muscles, the most inferior of the true tongue muscles, the **geniohyoid**, is also innervated by nerve C1 via the hypoglossal nerve, whereas the other three true tongue muscles—the **hyoglossus, genioglossus**, and **styloglossus**—are innervated by the hypoglossal nerve.

3.4.4.2 Tongue Muscles (CN XII and Cervical Nerve C1)

There are four true extrinsic tongue muscles (which are all hypobranchial muscles): Three of these muscles directly attach to the tongue and thus have "glossus" in their names (genioglossus, styloglossus, and hyoglossus); one of them does not (geniohyoid) (Plate 3.41). As noted previously, the other muscle that attaches onto the tongue and thus also has "glossus" in its name—the palatoglossus—is not a true tongue muscle but instead a pharyngeal (and thus branchial) muscle innervated by the vagus nerve. This muscle is another example of a structure whose anatomy and function in the adult differs from its developmental and evolutionary history. The **geniohyoid** muscle is deep (superior) to the mylohyoid, runs from the inferior mental spine of the mandible to the body of the hyoid bone, pulls the hyoid bone anteriorly, and is innervated by the **C1 nerve** that "hitchhikes" with the **hypoglossal nerve (CN XII)**, as explained above. The **genioglossus** is deep (superior) to the geniohyoid, runs from the superior mental spine of the mandible to the hyoid bone and tongue, protrudes (sticks out) the

**BOX 3.20 EASY WAYS TO REMEMBER:
FUNCTIONS OF THE INFRAHYOID MUSCLES**

To remember functions of the four infrahyoid muscles (Plate 3.19), think about which of the two structures connected by each of these muscles is "stronger," that is, harder and/or less moveable. As its name suggests, the more superficial of these muscles, the **sternohyoid muscle**, runs from the **sternum** to the **hyoid bone**. As the sternum is "stronger" than the hyoid bone (because it articulates with the ribcage and the clavicle, whereas the hyoid is held in place only by soft tissue), the sternohyoid muscle logically depresses the hyoid bone. Similarly, the **sternothyroid muscle** runs from the sternum to the **thyroid cartilage** and logically depresses the thyroid cartilage (because cartilage is "weaker," or softer, than bone). Conversely, the **omohyoid muscle** runs from the scapula (related to the name "*omo*" in Greek/Latin) to the hyoid bone and logically depresses the hyoid bone (the scapula is "stronger" than the hyoid bone because it is larger and attaches to the clavicle). Finally, the **thyrohyoid muscle**, which runs from the thyroid cartilage to the hyoid bone, logically elevates the thyroid cartilage because the bone is "stronger" than the cartilage.

tongue, and is innervated by the hypoglossal nerve, as are the styloglossus and the hyoglossus. As its name indicates, the **styloglossus** muscle runs from the styloid process to the lateral side of the tongue and thus retrudes (pulls in) and elevates the tongue. Lastly, the **hyoglossus** runs vertically from the body and greater horn of the hyoid bone to the side of the tongue and functions to depress and retrude the tongue. In addition to these four extrinsic muscles, the tongue includes four intrinsic muscles: **vertical tongue muscles**, **transverse tongue muscles**, **superior longitudinal muscles**, and **inferior longitudinal muscles**. Logically, all these intrinsic tongue muscles are also innervated by the hypoglossal nerve.

3.4.5 Extraocular Muscles (Eye Muscles, Nerves CN III, CN IV, and CN VI)

The extraocular eye muscles are shown in Plates 3.33 and 3.34. The **abducens nerve (CN VI)** only innervates the **lateral rectus** muscle, which exclusively laterally rotates (*abducts*) the eye. The **trochlear nerve (CN IV)** only innervates the **superior oblique muscle**, whose action is the opposite of what it first appears to be. The superior oblique muscle runs mainly superiorly and medially to the eye, so one might expect this muscle to elevate and adduct the eye. However, its tendon runs through a pulley-like sling, the *trochlea*, and turns posteriorly to insert on the posterior lateral surface of the eyeball; therefore, it *depresses* and *abducts* the eye. This makes sense because its function is not determined by the part near the origin that runs medially to the trochlea, but by the part near the insertion that runs posterolaterally to the eyeball. The other five extraocular muscles are innervated by the **oculomotor nerve (CN III)**. The **superior division of the oculomotor nerve** logically innervates the two more *superior* muscles—**levator palpebrae superioris** that elevates the eyelid (palpebrae = eyelids) and the **superior rectus** that elevates the eye, whereas the **inferior division of the oculomotor nerve** innervates the **inferior rectus** that depresses the eye, the **medial rectus** that abducts the eye, and the **inferior oblique** that elevates (this action is somewhat counter-intuitive as well) and abducts the eye because it runs posterolaterally to the posterolateral portion of the inferior surface of the eye. The actions of these muscles are fully integrated and actually more complicated enabling the eye to move rapidly yet smoothly while keeping horizontal visual field constant. These muscles contain unusual myosin isoforms to facilitate rapid movements.

BOX 3.21 EASY WAYS TO REMEMBER: FUNCTIONS OF THE EXTRAOCULAR MUSCLES

An easy way to remember the function of the six muscles attaching to the eye is to first remember that the superior oblique does the *opposite* of what would be expected based on the superior and medial position of its more proximal portion, as explained above: It actually depresses and abducts the eye. Then, based on this, you should remember "3-2-2-1," that is, 3 muscles abduct the eye, because this function is extremely important for the survival of animals, for instance to detect predators (these muscles are the lateral rectus, logically, and the two oblique muscles, that is the superior and inferior oblique); 2 muscles depress the eye (the inferior rectus logically and the superior oblique because this muscle does the opposite of what one would think it would do); 2 muscles elevate the eye (the superior rectus logically and the inferior oblique because this muscle elevates the eye rather than depressing it as one might think); 1 muscle only to adduct the eye (medial rectus, logically).

3.5 A Summary of the Head and Neck Musculoskeletal, Neurovascular, and Other Structures by Region

We believe that it is important for medical students to appreciate the skeletal, neurovascular, and muscular systems of the neck and head in their entirety, as presented in Sections 3.2 through 3.4, and also to understand the relationships among systems within each anatomical region, as emphasized in the following subsections. This book was written to allow either emphasis to be approached first, depending on individual preferences, or even on the type of study being done. For example, the more regional approach presented in this section—more similar to that given in "dissector" books—could be quickly used in the lab, whereas the more general, detailed approach given in the previous sections—more similar to that existing in many anatomy atlases—could be used to study in more detail, at home or in a library. Therefore this Section 3.5 presents a summarized description of the musculoskeletal and neurovascular structures of the head and neck already described in previous sections: These structures are now described in a topographical, regional context, and in some parts, the students are referred to the previous sections for more details. In contrast, some details given here, for instance regarding the eye or ear, are not given in the previous sections. Aside from these differences, the anatomy is the same, and as in the previous sections, we include in bold all the *terms* that medical students need to know, now included in their respective, specific topographic region.

3.5.1 Anterior Triangle of the Neck and Thyroid Gland

The anterior triangle of the neck (Plates 3.21 through 3.23 and 3.25) is bounded anteriorly by the median line of the neck, posteriorly by the anterior border of the **sternocleidomastoid muscle**, superiorly by the inferior border of the **mandible**, superficially by the investing layer of the deep cervical fascia (the "roof"), and deeply by the **larynx** and **pharynx** (the "floor") (Figure 3.6). The complex formed by the anterior and posterior **digastric muscles** and the **omohyoid muscle** divide the **anterior triangle of the neck** into four smaller triangles: the **muscular triangle**, **carotid triangle**, **submandibular triangle**, and **submental triangle**.

The contents of the **muscular triangle of the neck** are the **infrahyoid muscles**, the **thyroid gland**, and the **parathyroid glands**. The boundaries of this triangle are: superolateral, the superior belly of the omohyoid muscle; inferolateral, the anterior border of the sternocleidomastoid muscle; and

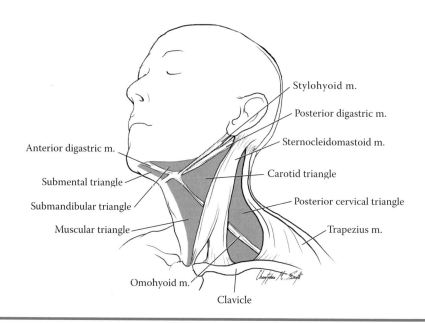

Stylohyoid m.

Posterior digastric m.

Sternocleidomastoid m.

Anterior digastric m.

Carotid triangle

Submental triangle

Submandibular triangle

Posterior cervical triangle

Muscular triangle

Trapezius m.

Omohyoid m.

Clavicle

Figure 3.6   Triangles of the neck, anterolateral view.

medial, the median plane of the neck. The infrahyoid muscles (Plate 3.19) were described in Section 3.4.4.1. The thyroid gland, located at vertebral levels C5-T1, has a **right thyroid lobe** and a **left thyroid lobe** connected by the **thyroid isthmus**. Frequently, the thyroid gland has a **pyramidal lobe** extending superiorly from the isthmus. On the posterior side of the thyroid gland lie the small **parathyroid glands**: usually two on each side of the parotid gland. The **superior thyroid artery** and the **inferior thyroid artery** are branches of the **external carotid artery** and of the **subclavian artery**, respectively (Plate 3.22; described in detail in Section 3.3.2). The **superior thyroid vein** and **middle thyroid vein** are tributaries of the **internal jugular vein**. The **inferior thyroid vein** is a tributary of the **brachiocephalic vein** (Plate 3.23). Posteriorly, the **recurrent laryngeal nerve** ascends close to the thyroid gland in the groove between the **trachea** and the **esophagus** (described in detail in Section 3.3.1.10).

The contents of the **carotid triangle of the neck** are: the **common carotid artery**, **internal carotid artery**, and the external carotid artery and its branches (see Section 3.3.2.1), the **hypoglossal nerve** (see Section 3.3.1.12), and the **internal laryngeal nerve** (that pierces the **thyrohyoid membrane**), and the **external laryngeal nerve** (that innervates the **inferior pharyngeal constrictor** and the **cricothyroid muscle**), and the **vagus nerve** (**CN X**) (see Section 3.3.1.10 and Box 3.13). The boundaries of this triangle are: inferomedial, the superior belly of the **omohyoid muscle**; inferolateral, the anterior border of the **sternocleidomastoid muscle**; and superior, the **posterior digastric muscle**. In the region of the carotid triangle, you can also observe the **inferior root of the ansa cervicalis** and the **superior root of the ansa cervicalis** (Plate 3.19; see Section 3.3.1.12) and the tributaries of the **internal jugular vein** (see Section 3.3.2.1).

The contents of the **submandibular triangle of the neck** are the **submandibular gland, facial artery, facial vein, stylohyoid muscle, hypoglossal nerve (CN XII)**, and **lymph nodes**. The boundaries of this triangle are: superior, the inferior border of the **mandible**; anteroinferior, the **anterior digastric muscle**; posteroinferior, the **posterior digastric muscle**; superficial (roof), the investing layer of deep cervical fascia; and deep, the **mylohyoid muscle** and **hyoglossus muscle**.

Lastly, the contents of the **submental triangle of the neck** are the **submental lymph nodes** and the boundaries of this triangle are: lateral, the right and left anterior digastric muscles; inferior, the **hyoid bone**; superficial (roof), the investing layer of deep cervical fascia; and deep, the **mylohyoid muscle** (Plate 3.25).

3.5.2 Posterior Triangle of the Neck

The **boundaries of the posterior triangle of the neck** (Plates 3.19 and 3.24) are: anterior, the posterior border of the **sternocleidomastoid muscle**; posterior, the anterior border of the **trapezius muscle**; inferior, the middle one-third of the **clavicle**; superficial (roof), the investing layer of the **deep cervical fascia**; and deep (floor), the muscles of the neck covered by **prevertebral fascia** (Figure 3.6). The contents of this triangle include the **accessory nerve (CN XI)** (see Section 3.3.1.11). In the superficial region of the neck, you can observe the **anterior jugular vein** and **external jugular vein** (Plate 3.25; described in Section 3.3.2.2), the **platysma muscle**, and the **cutaneous nerves of the cervical plexus** (described in Section 3.3.1.12).

3.5.3 Root of the Neck

The **boundaries of the root (base) of the neck** (Plates 3.20 and 3.22 through 3.25) are: anterior—the **manubrium of the sternum**; lateral—the 1st pair of **ribs**; and posterior—body of **vertebra T1**. The root of the neck lies superiorly to the **superior thoracic aperture**. The contents of the root of the neck include: the **subclavian vein, external jugular vein**, and **bra-chiocephalic vein**, the **subclavian artery** and its branches (see Section 3.3.2), the **thoracic duct, right lymphatic duct, phrenic nerve**, cervical portion of the **sympathetic trunk**, the **vagus nerve** and the **recurrent laryngeal nerves** (see detailed description in Section 3.3.1.10). Remember that the **subclavian vein, phrenic nerve, suprascapular artery**, and **transverse cervical artery** are superficial to the **anterior scalene mus-cle**, whereas the **subclavian artery** and **roots of the brachial plexus** are deep to this muscle, passing between it and the **middle scalene muscle**, in the **interscalene triangle**. The anterior scalene muscle is also the major landmark for the root of the neck, in particular to define the three parts of the subclavian artery (see Section 3.3.2.1). The anterior scalene, **middle sca-lene**, and **posterior scalene** are innervated exclusively by cervical nerves, so they are not true head muscles in the evolutionary/developmental sense.

3.5.4 Face, Scalp, and Parotid Region

As explained in Section 3.3.1.5, the **skin** of the face receives sensory inner-vation from the **trigeminal nerve (CN V)**, which has three branches (divi-sions) (Plate 3.15). First, the **ophthalmic nerve (CN V₁)** gives rise to the **supraorbital nerve, supratrochlear nerve**, and **lacrimal nerve** that pass through the **orbit** and innervate the skin of the **forehead, upper eyelids**, and **nose**. Second, the **maxillary nerve (CN V₂)** gives rise to the **infraor-bital nerve** that emerges from the **infraorbital foramen** and innervates the **lower eyelids** and medial **cheek**. Third, the **mandibular nerve (CN V₃)** gives rise to the **mental nerve** that innervates the skin of the **lower face**—including the **chin** and **lower lip**—the **buccal nerve** (that inner-vates the cheek), and the **auriculotemporal nerve** that innervates part of the side of the head. As explained in Section 3.3.1.12, the back of the head and the area around the ear receive sensory innervation from **spinal cervi-cal nerves 2 and 3**.

The **muscles of facial expression** are innervated by the **facial nerve (CN VII)** (Plates 3.16 and 3.26; described in detail in Section 3.3.1.7). The facial nerve exits the skull through the **stylomastoid foramen**, then enters the **parotid gland** giving off five branches: the **temporal branch, zygo-matic branch, buccal branch, mandibular (or marginal mandibular) branch**, and **cervical branch** (Plate 3.15). The muscles of facial expres-sion (Plate 3.27; described in detail in Section 3.4.2) include the **platysma muscle** which covers the anterior neck; the **orbicularis oculi** (made up of the **orbital part, palpebral part**, and **lacrimal part** which encircles the **palpebral fissure**, or eyelid). They also include the muscles which surround the mouth (**orbicularis oris**) and attach to the lips (**levator labii superioris, levator labii superioris alaeque nasi, depressor labii inferioris, levator anguli oris, depressor anguli oris**, and **risorius**), and the muscles of the forehead and scalp (**procerus** and **occipitofrontalis**). The occipitofrontalis is made up of two muscles (occipitalis and frontalis) connected by the **epicranial aponeurosis** sandwiched between the superficial and deep layers of **skin** and **connective tissue** of the **scalp**. Lastly, the muscles of facial expression also include the muscles of the cheek and cheekbone (**buccinator**, which attaches to the **pterygomandibular raphe**, and **zygomaticus major** and **zygomaticus minor**), of the chin (**mentalis**), and of the external ear: the **auricularis anterior, auricularis posterior, auricularis superior**, and the **temporoparietalis**.

The main vessels of the face are the **facial artery, inferior labial artery, superior labial artery**, and **angular artery**, as well as the **facial vein** and **angular vein** (Plates 3.22 and 3.23; described in detail in Section 3.3.2).

The boundaries of the **parotid region (parotid bed)** are: posterior—mastoid process and **posterior digastric muscle**; anterior—**medial pterygoid muscle**, mandibular ramus, and **masseter muscle**; medial—styloid process and associated muscles (**stylopharyngeus, styloglossus**, and **stylohyoid muscle**); and posterosuperior—floor of the external acoustic meatus. The **parotid bed** is occupied by the **parotid gland** and the nerves and vessels that pass through it (Plate 3.26). The parotid gland develops as an evagination of the **oral mucosa**, and in adults is enclosed within the **parotid sheath** and connected to the **parotid duct**. The parotid duct pierces the **buccinator muscle**, anterior to the masseter muscle, and deep to the **buccal fat pad**. Superior to the parotid duct is the small **transverse facial artery** (Plate 3.22). The bony features of the parotid region include the **external acoustic meatus, styloid process, stylomastoid foramen**, and **mastoid process** of the **temporal bone**, and the **mandibular fossa, mandibular head, mandibular neck, angle of the mandible**, and **ramus** of the mandible (Plate 3.8; described in Section 3.2). Within the parotid gland lie the branches of the **facial nerve** (Plate 3.26; described in Section 3.3.1.7), and the **auriculotemporal nerve**, which carries postganglionic parasympathetic nerve fibers from the **otic ganglion** to the **parotid gland** and are described in more detail in Section 3.3.1.5 (Plate 3.17). The vessels of the parotid region are the **external jugular vein, maxillary** vein, **superficial temporal vein**, **external carotid artery**, **maxillary artery**, and **superficial temporal artery** (Plates 3.26, 3.22, and 3.23; described in detail in Section 3.3.2).

3.5.5 Temporal Region and Mandible

The **temporal region** (Plates 3.8 and 3.28 through 3.30) includes the **temporal fossa** and the **infratemporal fossa**, which lie superiorly and inferiorly to the zygomatic arch, respectively. The bony features of this region are shown in Plate 3.8 and described in detail in Section 3.2. Briefly, the **parietal bone** and **frontal bone** include the **superior and inferior temporal lines** for attachment of the temporalis muscle. The **zygomatic arch** is formed by the **zygomatic process of the temporal bone** and the **temporal process of the zygomatic bone**. The temporal fossa is formed by the parietal, **frontal, squamous part of the temporal**, and **greater wing of the sphenoid**, and contains the **temporalis muscle**. The infratemporal fossa contains the **medial pterygoid muscle, lateral pterygoid muscle**, branches of the **mandibular nerve (CN V$_3$)** and the **maxillary artery**, and the **venous pterygoid plexus** converging into the **maxillary veins** (Plate 3.29). These two fossae communicate with each other through the interval between the **zygomatic arch** and the lateral surface of the skull. The **pterygomaxillary fissure** lies between the **lateral plate of the pterygoid process** of the **sphenoid bone** and the **infratemporal surface of the maxilla**. The **pterygopalatine fossa** lies at the superior end of the pterygomaxillary fissure, and the **sphenopalatine foramen** (which opens into the **nasal cavity**) is medial to the fossa (Plate 3.8c; see Box 3.9 for an easy way to remember these structures). The **inferior orbital fissure** lies between the **maxilla** and the **greater wing of the sphenoid bone**, which contains the **foramen ovale** and **foramen spinosum**.

The **mandible** (Plate 3.8b; described in detail in Section 3.2.2) includes the **mandibular body, mandibular angle, mandibular ramus, mandibular notch, coronoid process, mandibular neck, mental foramen**, and **mandibular head** which articulates with the temporal bone through the **temporomandibular joint** (Plate 3.28). This joint, formed by the mandibular head and the **articular eminence (tubercle) of the temporal bone**, includes the **superior synovial cavity, inferior synovial cavity, articular disc**, and **capsule of the temporomandibular joint**, reinforced by the **temporomandibular ligament**. On the **inner aspect of the mandible** lie the **digastric fossa**, the **superior and inferior mental spines**, the **mylohyoid line**, the **lingula** for the attachment of the **sphenomandibular ligament**, and the **submandibular fossa**.

The **boundaries of the infratemporal fossa** (Plate 3.29) are: lateral—mandibular ramus; anterior—infratemporal surface of maxilla; medial—lateral plate of pterygoid process; roof—greater wing of sphenoid bone. The **temporalis**—and associated **temporal fascia** (which is deep to the **superficial temporal vessels** and the **auriculotemporal nerve**)—and **masseter** are the most superficial (lateral) muscles of the temporal region. The **lateral pterygoid muscle** and **medial pterygoid muscles** are deep to the mandible; these are all muscles of the 1st arch, described in Section 3.4.1.

The **inferior alveolar nerve** is a branch of the **mandibular nerve** (**CN V₃**) (Plate 3.15; described in detail in Section 3.3.1.5). Its course takes it through the three foramina/canals: first the **mandibular foramen**, then the **mandibular canal**, where it sends branches to the **mandibular teeth**, and finally the **mental foramen** where it becomes the **mental nerve**. The **mylohyoid nerve** is a branch of the mandibular nerve that runs in the **mylohyoid groove**. Another branch is the **lingual nerve**, which is joined by the **chorda tympani** before entering the oral region. Other branches of the mandibular nerve that lie in the temporal region are the **buccal nerve** and the **auriculotemporal nerve** (which supplies the **dura mater**), also described in Section 3.3.1.5.

The external carotid artery divides, superiorly, into the **maxillary artery** and **superficial temporal artery**, which lie in the temporal region (Plates 3.22 and 3.29; described in detail in Section 3.3.2.1). The main branches of the maxillary artery are the **middle meningeal artery, anterior and posterior deep temporal arteries, masseteric artery**, running superiorly, the **inferior alveolar artery**, running inferiorly to enter the **mandibular foramen** with the **inferior alveolar nerve**, and the **buccal artery**, running anteriorly and inferiorly. After giving rise to these branches, the maxillary artery runs anteriorly toward the **pterygopalatine fossa** and divides into four terminal branches: the **posterior superior alveolar artery, infraorbital artery, descending palatine artery**, and **sphenopalatine artery**.

3.5.6 Craniovertebral Joints and Retropharyngeal Space

The craniovertebral joints lie between the occipital bone and the first two cervical vertebrae (Plate 3.31): the **atlas** (vertebra C1)—including the **anterior**

arch, superior articular facet, transverse process, and **posterior arch**—and the **axis** (vertebra C2; its name implying a point of rotation)—including the **dens** (see Section 6.1 and Box 6.2). The dens (meaning tooth-like) is an elongated median process that projects into the spinal canal of the atlas, allowing it to rotate relative to the axis (i.e., this is the "no" joint). The dens (which, developmentally, corresponds to the body of C1) is held tightly against the floor of the spinal canal in the atlas by the **transverse ligament of the atlas**, so it does not impinge on the **spinal cord**. The **occipital bone**—including the **occipital condyle, pharyngeal tubercle**, and **foramen magnum**—are shown in Plate 3.10. The **atlanto-occipital joint** is made by the occipital condyle and the superior articular facet of the atlas.

The **retropharyngeal space** is situated posterior to the cervical viscera and anterior to the vertebrae. It extends from the **basilar part of the occipital bone** to the **superior mediastinum** and forms a dangerous route for the spread of infection from the pharyngeal wall down to the thorax. Lateral to this space is the **sympathetic trunk**. Along the trunk, swellings of the **superior, middle, and inferior cervical sympathetic ganglia** are found (often the inferior cervical ganglion and the **1st thoracic ganglion** fuse into one **stellate ganglion**, or **cervicothoracic ganglion**). Gray **rami communicantes** connect these ganglia to the cervical spinal nerves. Superiorly, the **internal carotid nerve** carries postganglionic fibers from the superior cervical ganglion to all the structures of the head (Plate 3.20). The **prevertebral fascia** covers the **prevertebral muscles (longus colli** and **longus capitis)** and the **lateral vertebral muscles (anterior, middle, and posterior scalene muscles)**.

3.5.7 Orbit

BOX 3.22 DEVELOPMENT OF THE EYE: TWO COMPARTMENTS AND A MUSCULOSKELETAL APPARATUS

The primordia of both eyes are first biochemically detectable as a single area in the midline of the rostral neural plate, which is the flat epithelial sheet that folds to form the dorsal neural tube. In response to signals from underlying prechordal mesoderm, this single **eye field** becomes divided into separate left and right eye primordia. As the rostral **neural plate** folds to form the **prosencephalon**, the paired **optic primordia** become located on the lateral walls of the **diencephalon**. They then evaginate (bulge outwards) to form the **optic vesicles** (Plate 3.3a and b), which project laterally from the wall of the diencephalon. These vesicles extend up against the lateral body wall. Subsequently, the anterior (outermost) surface of each optic vesicle invaginates (collapses inwards) bringing the original outer wall of the vesicle into close apposition of the former inner surface. This now is the **optic cup**. Within each optic cup, the newly formed inner layer forms the **neural retina** whereas the now superficial layer becomes the **pigmented retinal epithelium**. The original connection with the diencephalon, the **optic stalk**, constricts and becomes occluded.

At the site of contact between the optic vesicle and surface **ectoderm**, the surface **epithelium** thickens to form a **lens placode**. While the **optic vesicle** is forming an optic cup, this placode invaginates, separates from the overlying epithelium, and closes to form the **lens vesicle**. As the **lens** grows, its original lumen becomes obliterated and its anterior epithelial margin thins. In contrast, cells in the posterior part of the lens become elongated in the anterior–posterior direction and tightly packed. They lose most of their cytoplasmic contents, including their nuclei, and become dehydrated. This cellular remodeling renders the lens transparent.

Neurogenesis in the **retina** begins centrally, close to the optic stalk, and progresses peripherally toward the lips of the optic cup. Within the retina, the transducing receptors—**rods** and **cones**—come to lie superficially, adjacent to the pigmented layer, whereas the principle output neurons—**retinal ganglion neurons**—lie adjacent to the vitreous fluid within the optic cup. Axons from retinal ganglion neurons project toward the site of optic stalk attachment, but there they exit the eye as the **optic nerve** (**CN II**) (Plate 3.14) and travel medially along a groove on the ventral surface of the **optic stalk**. On reaching the **optic chiasm**, beneath the diencephalon, CN II axons either cross the midline or remain on the same side, depending on the location on the retina from which they originated, and then reenter the CNS.

The anterior margin of the optic cup, where pigmented and neural retina layers meet, expands and partially overgrows the anterior surface of the lens. This bilayered epithelial extension, together with some overlying neural crest cells, forms the **iris** and, later, the **ciliary body** (Plate 3.3b). Deposits of pigment granules within iris tissue determine the "color" of the eye; circular and radial muscles in the iris facilitate constriction or dilation of the pupil. The ciliary body and associated muscles attach to the equatorial zone of the lens and aid in visual accommodation by modulating tension applied to the lens.

The optic vesicle becomes surrounded by neural crest cells except along its anterior surface, where its contact with the lens placode blocks crest cell immigration. This population of **periocular mesenchyme** is invaded by **angioblasts** that form the **choroid layer**, which is the vascular plexus on the outer surface of the eye. Other blood vessels penetrate the eye along the groove of the optic stalk to form the **hyaloid vascular plexus**. These intraocular blood vessels ramify (branch) throughout the optic vesicle. Later in development, only those intimately associated with the neural retina will remain. Periocular mesenchyme is also invaded by myoblasts from paraxial mesoderm; these form the **extraocular muscles** (see also Box 3.2). **Neural crest cells** form all periocular connective tissues including the **sclera**, a fibrous connective tissue onto which the extraocular muscles attach (Plate 3.32).

Crest cells do not surround the anterior surface of the eye until after the lens has formed. They then invade the space between the lens and overlying surface ectoderm. Here they form the deep epithelial layer, the posterior epithelium (**corneal endothelium**) of the **cornea**, and secondarily invade the matrix between this and the overlying surface ectoderm to form the thicker **stromal layer** of the cornea. The corneal stroma is composed of layers of keratinocytes that are precisely aligned so as to render the cornea transparent; the matrix of the stroma also prevents ingression of angioblasts and thereby maintains the necessary corneal avascularity. The **eyelids (palpebrae)** (Plate 3.32) develop from mesenchymal cells that proliferate in bands around the cornea, forming folds that cover the anterior surface of the eye. The palpebrae elongate and contact each other, becoming tightly adherent during most of gestation. As the eyelids are developing, local invaginations from their inner surface form cords that branch and hollow out. These cords become ducts and acini (terminal sacs of secretory cells) associated with the **lacrimal gland** (Plate 3.14). The eyelids remain tightly closed during most of gestation to protect the surface of the cornea from exposure to amniotic fluid.

The bones that form the adult human **orbit** are lined with periosteum, which is specifically named the **periorbita** in this region. The orbital contributions from the **frontal bone** include the **supraorbital notch** and **orbital surface** superiorly and **lacrimal fossa** medially. Orbital portions of the sphenoid include the **optic canal, lesser wing, greater wing**, and **superior orbital fissure**, all located posteriorly in the orbit. The **zygomatic bone** makes up the lateral wall, whereas the **ethmoid bone**, the **posterior lacrimal crest**, and **lacrimal groove** of the **lacrimal bone** (including the **fossa for lacrimal sac**) make up the medial wall. The **maxillary bone** including the **anterior lacrimal crest, infraorbital groove, and infraorbital foramen** make up the floor of the orbit (Plates 3.7 and 3.8; described in detail in Section 3.2). The orbit communicates with the **anterior cranial fossa** (superiorly), maxillary sinus (inferiorly), and **ethmoidal air cells** (medially).

The external anatomy of the orbital region (Plate 3.32) includes the **eyelashes (cilia), palpebral fissure** (opening between **eyelids**, or **rima**), **medial and lateral palpebral commissures** (joining of upper and lower eyelids), **medial and lateral angles (canthi)** of the eye, **lacrimal caruncle** bump and the **lacrimal lake** surrounding it, the **lacrimal papilla** bump and the **lacrimal puncta** opening at its apex. The anterior aspect of the eyeball includes the **sclera, cornea, iris**, and **pupil**. The sclera and lids are lined by the **bulbar conjunctiva** and **palpebral conjunctiva**, forming the **superior and inferior conjunctival fornices** and **conjunctival sac of the eye**.

The **tarsal glands** are embedded in the posterior surface of each **tarsal plate**, the fibrous "skeleton" of the upper eyelid. The **orbital septum** is a sheet of connective tissue separating the superficial facial fascia and the contents of the orbit. The tarsal glands drain by orifices lying posteriorly to the eyelashes and secrete an oily substance onto the margin of the eyelids that prevents the overflow of tears (lacrimal fluid). The **lacrimal gland** lies in the **lacrimal fossa** of the frontal bone. The **lacrimal sac** lies posterior to the **medial palpebral ligament**, which is attached to the **anterior lacrimal crest** that forms the anterior border of the **lacrimal groove** (Plate 3.32). The lacrimal sac receives lacrimal fluid from the medial angle of the eye through the **lacrimal canaliculi**. When lacrimal fluid accumulates in excess and cannot be removed from the medial corner of the eye via the lacrimal canaliculi, it overflows the eyelids (visible crying or shedding of tears). Tears also drain into the nasal cavity via the lacrimal sac, resulting in a runny nose.

Three main sensory nerves branch from the **ophthalmic nerve (CN V$_1$)** and lie in the orbital region: the **frontal nerve**, the **lacrimal nerve**, and the **nasociliary nerve** (Plates 3.15 and 3.33; described in detail in Section 3.3.1.5). The frontal nerve gives rise to the **supratrochlear nerve** and the

supraorbital nerve. The nasociliary nerve gives rise to the **long ciliary nerves**, **anterior ethmoidal nerve** (which leaves the orbit and enters the nasal region, giving rise to the **external nasal nerve**), and **posterior ethmoidal nerve**.

The **oculomotor nerve (CN III)** (Plates 3.15 and 3.33; described in detail in Section 3.3.1.3) carries with it preganglionic parasympathetic fibers. These fibers synapse on cells in the **ciliary ganglion**, which is found between the **optic nerve (CN II)** and the lateral rectus muscle. The postganglionic fibers connect to the eyeball **via short ciliary nerves** and project to the smooth muscles of the sphincter pupillae and the **ciliary body** (Plate 3.20). Some sensory fibers from the nasociliary nerve and some postganglionic sympathetic fibers also pass through the ciliary ganglion on their way to the eyeball, but they do not synapse there. The **lacrimal nerve** also carries with it some parasympathetic fibers, but these are postganglionic fibers from the **pterygopalatine ganglion**, which receives input from the **greater petrosal branch of the facial nerve**. These fibers activate the lacrimal gland to produce tears.

The muscles of the orbit are the extraocular muscles: the **superior rectus**, **inferior rectus medial rectus**, **lateral rectus**, **superior oblique**, and **inferior oblique**, all of which originate from the **common tendinous ring** and insert on the eyeball, and the **levator palpebrae superioris** which opens the eyelid. The anatomy, innervation, and function of the extraocular muscles (Plates 3.33 and 3.34) are described in Section 3.4.5; see also Box 3.21 for easy ways to memorize these muscles and their functions.

Most of the blood vessels of the orbital region (Plate 3.34) are branches of the **ophthalmic artery**, which enters the orbit through the optic canal, inferior to the optic nerve. It gives off the **central artery of the retina**, then divides into the lacrimal artery, the supratrochlear and supraorbital arteries (Plate 3.22). The major veins of the orbital region are the **superior ophthalmic vein**, which anastomoses with the **angular vein**, and the **inferior ophthalmic vein** (Plate 3.23; described in detail in Section 3.3.2).

3.5.8 Pharynx

In human adult anatomy, the term **pharynx** is usually used to refer to the region that extends from the level of the base of the skull to the inferior border of the cricoid cartilage (vertebral level C6) anterior to the retropharyngeal space (Plates 3.35 through 3.37). The three layers of the **pharyngeal wall** are the **buccopharyngeal fascia**, the **muscular layer**, and **pharyngobasilar**

fascia. The muscular layer is composed of an outer circular part (**superior, middle, and inferior pharyngeal constrictors**) and an inner longitudinal part (**salpingopharyngeus** and **palatopharyngeus**, fused inferiorly with the **stylopharyngeus**). Attachments of the pharyngeal constrictor muscles include: the **pharyngeal raphe** in the posterior midline (all 3 muscles); the **pharyngeal tubercle of the occipital bone** superiorly, the **pterygomandibular raphe** anterosuperiorly, the tongue and the mandible (the superior constrictor); the hyoid bone (the middle constrictor), and the thyroid and cricoid cartilages inferiorly (the inferior constrictor). The dense connective tissue membrane that attaches the superior margin of the superior constrictor to the skull is the **pharyngobasilar fascia**. The **stylopharyngeus** muscle and the **glossopharyngeal nerve (CN IX)** pass between the middle and superior constrictors (Plates 3.21 and 3.36; described in detail in Section 3.4.3.1). The **pharyngeal plexus of nerves** is located on the posterolateral surface of the pharynx and receives branches of the glossopharyngeal nerve—sensory to the pharyngeal mucosa, **vagus nerve (CN X)**—motor to almost all pharyngeal muscles, and contributing to the pharyngeal plexus via the **pharyngeal plexus of the vagus nerve**, and **superior cervical sympathetic ganglion**—vasomotor (Plates 3.20 and 3.22).

The internal aspect of the pharynx is shown in Plates 3.36 and 3.37. The **nasopharynx** lies posterior to the nasal cavity, and specifically to the **choana**, which is the transition region from the **nasal cavity** to the nasopharynx. The choanae of the two sides are separated by the posterior end of the **nasal septum** (Plate 3.38). The nasopharynx includes the **opening of the pharyngotympanic tube (auditory tube** or **Eustachian tube**, connecting the nasopharynx to the **tympanic cavity**, also called the **middle ear cavity**), the **torus tubarius** (cartilage of the auditory tube that is covered by mucosa), the **salpingopalatine fold** (mucous membrane containing the **levator veli palatini muscle**), the **salpingopharyngeal fold** (mucous membrane containing the **salpingopharyngeal muscle**), the pharyngeal recess, and the **pharyngeal tonsil (adenoid)**, that is located in the mucous membrane of this recess.

The **oropharynx** is posterior to the **oral cavity**, lying inferiorly to the **soft palate** and superiorly to the **epiglottis**. The **palatoglossal fold** (mucous membrane containing the **palatoglossus muscle**) forms a dividing line between the oral cavity and the oropharynx, and the transitional region between the right and left palatoglossal folds is called the **fauces**. Between the palatoglossal fold and the palatopharyngeal fold (mucous membrane containing the **palatopharyngeus muscle**) is the **palatine tonsil** (Plate 3.36).

The **laryngopharynx** lies posterior to the larynx, extending from the epiglottis to the lower border of the cricoid cartilage. It includes the **inlet (aditus) of the larynx** and the **piriform recess**, which is bordered by the **larynx** (medially), **thyroid cartilage** (laterally), and **inferior pharyngeal constrictor muscle** (posteriorly) (Plate 3.37). The **piriform fossa** is a common place for ingested items to become trapped. Underlying the mucosa is the **internal laryngeal nerve** that provides sensory innervation above the **vocal folds**. Attempts to remove foreign objects or irritation from trapped medications (e.g., an aspirin) may cause damage to this nerve and compromise the choking reflex where it is most important.

3.5.9 Nose and Nasal Cavity

As explained in detail in Section 3.2, the **nasal cavity** is bounded laterally by the **lateral nasal wall** composed of parts of the maxilla, **ethmoid bone, lacrimal bone, sphenoid bone (body, medial plate of pterygoid process, lateral plate of pterygoid process)**, and the **perpendicular plate** of the **palatine bone**. Superiorly the nasal cavity is bounded by part of the nasal septum and parts of the **nasal bone, cribriform plate** of ethmoid bone, and sphenoid bone, and inferiorly by the hard palate, made up of the palatine process of the maxilla, and **horizontal plate** of the palatine bone (Plate 3.9). The nasal cavity is also divided in the midline by the **nasal septum**, made up of the **septal cartilage** (including the **lateral nasal cartilage**), the **perpendicular plate of the ethmoid bone**, and the **vomer**. On the lateral nasal wall are the nasal conchae (**superior nasal concha, middle nasal concha**, and **inferior nasal concha**). The lateral nasal wall is pierced by a number of foramina and channels, including the **sphenoethmoidal recess, superior meatus** (containing the **opening of the posterior ethmoidal cells**), **middle meatus, inferior meatus, semilunar hiatus** (with openings of the **frontal sinus, anterior ethmoidal cells**, and **maxillary sinus**), and openings for the **nasolacrimal duct** and **sphenoidal sinus** (Plate 3.38). The middle ethmoid cells are contained in the **ethmoid bulla**, and they open on or near this bony feature. The **vestibule** (main part of the nasal cavity) is anterior to the inferior meatus, while the **atrium** (smaller region that opens to the external nares formed by the **alar cartilage**) is anterior to the inferior meatus. In the mucosa of the nasal septum lie the **nasopalatine nerve** and **sphenopalatine artery**, which leave the nasal cavity through the **incisive foramen** to enter the palate (Plates 3.29 and 3.39).

3.5.10 Hard Palate and Soft Palate

The hard palate forms the anterior two-thirds of the palate, whereas the soft palate forms the posterior one-third (Plate 3.36). As explained in detail in Section 3.2, the **hard palate** is made up of the **horizontal plate of the palatine bone** and the **palatine process of the maxilla**. The hard palate is pierced by several foramina: the **incisive foramen, greater palatine foramen**, and **lesser palatine foramen**. The soft palate includes the palatine glands and muscles such as the **musculus uvulae, levator veli palatini**, and the aponeurosis of the **tensor veli palatini**. When one removes the pharyngeal mucosa (see Section 3.5.8), one can then see the **palatopharyngeus** muscle running from the hard palate and palatine aponeurosis to the thyroid cartilage and pharyngeal wall. The palatopharyngeus is innervated by the vagus nerve (CN X), elevates the larynx during swallowing, and forms part of the **longitudinal layer of the pharynx** together with the **salpingopharyngeus**, and the **stylopharyngeus**. The salpingopharyngeus is also innervated by the vagus nerve (CN X) and runs from the cartilage of the auditory tube and blends with the palatopharyngeus to attach distally onto the thyroid cartilage and pharyngeal wall, thus also contributing to the elevation of the larynx during swallowing. The **levator veli palatini** is also innervated by the vagus nerve (CN X), and runs from the cartilage of the auditory tube and adjacent temporal bone to the palatine aponeurosis of the soft palate, thus being able to elevate the soft palate. The palatoglossus is also innervated by the vagus nerve (CN X); and contrary to what its name indicates, is not a true tongue (and thus hypobranchial) muscle (see Section 3.4). It runs from the palatine aponeurosis to the lateral side of the tongue, and thus elevates the tongue and depresses the soft palate. The **glossopharyngeal nerve** innervates the stylopharyngeus (see Section 3.4) and provides sensory innervation to the mucosa of the posterior one-third of the tongue and the posterior wall of the pharynx. The **tensor veli palatini** (which originates from the **scaphoid fossa**) is not a pharyngeal muscle in the sense that it is instead a 1st arch (mandibular) muscle that is accordingly innervated by the trigeminal nerve (CN V). Therefore, it makes sense that this muscle is more superficial (lateral) than true pharyngeal muscles such as the levator veli palatini. The tensor veli palatini originates from the scaphoid fossa of the sphenoid bone and runs vertically between the **pterygoid plates**. Its tendon wraps around the **hamulus of the medial pterygoid plate** to spread out in a horizontal plane as the palatine aponeurosis and merges with the fibers from the contralateral side. When contracted, it tenses the soft palate.

During swallowing, the tensor veli palatini and the levator veli palatini contract together to seal the soft palate against the posterior pharyngeal wall and prevent food and liquids from entering the nasal pharynx.

Remember that the **sphenopalatine artery** runs with the **nasopalatine nerve** from the **sphenopalatine foramen** to the **incisive canal**. The sphenopalatine artery branches into the **posterior lateral nasal artery** and the **posterior septal branch** (Plate 3.29; described in detail in Section 3.3.2.1). The **descending palatine artery** is a branch of the **maxillary artery** that runs through the **greater palatine canal** before bifurcating into the **greater palatine artery** and the **lesser palatine artery**.

The greater and lesser palatine nerves supply both the hard and soft palates and carry with them postganglionic parasympathetic fibers. Remember that **the nerve of the pterygoid canal** (**Vidian nerve**) enters the pterygopalatine fossa, is connected anteriorly to the **pterygopalatine ganglion**, and carries preganglionic parasympathetic axons from the **greater petrosal nerve** and postganglionic sympathetic axons from the **deep petrosal nerve** (Plates 3.20 and 3.39; described in detail in Section 3.3.1.5). The lesser palatine nerve and the greater palatine nerve carry these fibers from the pterygopalatine ganglion through the greater palatine canal and pass through the lesser and greater palatine foramina, respectively.

3.5.11 *Oral Region*

BOX 3.23 DEVELOPMENT OF THE TONGUE, CHEEKS, AND LIPS

The **tongue** forms initially as several swellings that develop in the floor of the **pharynx** and **stomodeum**, which form by folding of the surface epithelium when the brain elongates and bends ventrally (flexion) at the level of the future midbrain (Figure 3.7). Thus, the surface of the tongue contains both endodermal and ectodermal **epithelia. Neural crest cells** from several branchial arches form the connective tissues of the tongue. The most rostral swellings are a pair of **distal** (**lateral**) **tongue swellings**. Caudal to these are three unpaired swellings, the **median tongue swelling**, **proximal tongue swelling**, and the **epiglottal swelling**, which forms the **epiglottis** (Plate 3.40). The **basihyoid cartilage** is a 2nd branchial arch structure that develops in the midline within the proximal tongue swelling; it functions to stabilize the root of the tongue and to provide muscle attachment sites for both extrinsic and intrinsic tongue muscles (see Section 3.4.4).

All intrinsic and extrinsic tongue muscles are innervated by the **hypoglossal nerve** (CN XII), whose cell bodies are in the caudal hindbrain—a long way caudal to other suggested "1st arch" structures (Plate 3.2). Mapping studies in chickens have shown that all tongue muscles arise from myotomes of somites 2–5, which are also located beside the caudal hindbrain. These cells move as a band called the **hypoglossal cord** ventrally from the **somites**, around the **pharynx**, then cranially to invade the tongue swellings; the **infrahyoid muscles** are formed in a similar way (see Section 3.4.4). Some recent developmental studies in mice (from members of Carmen Birchmeier's lab) suggest that part of the extrinsic tongue musculature has a branchiomeric (not somitic) origin, being therefore derived from **cranial mesoderm**. However, other similar studies in mice continue to support an exclusive somitic origin of these muscles (Plate 3.2).

Growth of the tongue is very disproportional. The distal swellings expand to form most of the **body of the tongue** (Plate 3.40).

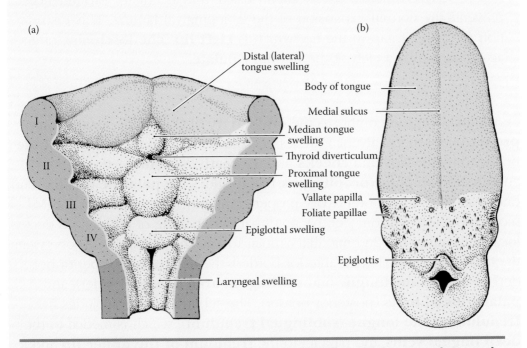

(a)

(b)

Distal (lateral) tongue swelling

Body of tongue

Medial sulcus

Median tongue swelling

Thyroid diverticulum

Proximal tongue swelling

Vallate papilla

Foliate papillae

Epiglottal swelling

Epiglottis

Laryngeal swelling

I

II

III

IV

Figure 3.7   (a and b) Schematic dorsal views of the floor of the stomodeum and pharynx, illustrating the locations of swellings that form the tongue and epiglottis. Light blue-ectodermal epithelium and yellow-endodermal epithelium.

Consequently, most of the surface of the tongue is innervated by the **mandibular nerve (CN V₃)**. Specialized **chemoreceptors** and associated epithelial components of **taste buds** on the surface of the tongue are innervated by branches of the **facial nerve** (CN VII) and of the **glossopharyngeal nerve** (CN IX).

Although the maxillary and mandibular prominences (Plate 3.2) elongate independently, they remain connected by a mesenchymal bridge that is lined internally by stomodeal epithelium and externally by surface **ectoderm**. Initially there are no cheek pouches; rather, the early embryonic oral cavity has solid lateral walls. An epithelial thickening, the **labiogingival lamina**, forms along the apposed surfaces of both prominences. This thickened epithelial ridge invaginates and then hollows out to form a **labiogingival groove**, which expands dorsally and ventrally to establish the **oral vestibule** that separates the **lips** and **cheek** from **teeth** and deeper skeletal structures. The lamina extends along the margins of the mandibular and maxillary prominences and also across the midline along the frontonasal prominence. Any disruption of the proper fusion between medial nasal, lateral nasal, and maxillary prominences prevents the normal expansion of the labiogingival lamina. This disruption results in a gap in the lip, which is **cleft lip (cheiloschisis)**, often accompanied by a cleft of the **primary palate**.

The **oral region** comprises the **oral cavity** (including the **oral cavity proper** between the alveolar arches and the **teeth**, and the **oral vestibule** bounded by the **lips** and **cheeks** and the **teeth** and **gums**), the **palate**, and the portion of the **oropharynx** that includes the **palatine tonsils**. The oral vestibule and the oral cavity proper communicate posterior to the 3rd molar tooth (or "wisdom" tooth); note that the opening of the parotid duct is located lateral to the 2nd maxillary molar tooth. The oral cavity lies between the teeth and gums (lateral/anterior border), the hard palate (superior border), mucosa of the **tongue** and **sublingual area** (inferior border), and **palatoglossal folds** (posterior border). The sublingual area includes the **frenulum of the tongue (sublingual frenulum)**, which is medial to the **deep lingual veins** and parallel to the **frenulum of the upper lip** and to the **frenulum of the lower lip**.

Other structures of the oral cavity include the **sublingual fold (plica sublingualis)** covering the **sublingual gland** and the **opening of**

the submandibular duct in the **sublingual caruncle**. The sublingual gland has about 12 small ducts that drain along the summit of the sublingual fold, but are not visible to the naked eye. The **submandibular gland** wraps around the posterior edge of the **mylohyoid muscle**, with most of it filling the **submandibular triangle of the neck** (Plate 3.25). The deep lobe lies deep to the mylohyoid and gives rise to the submandibular duct which runs along the medial edge of the sublingual gland. You may be able to watch this particular salivary gland in action if, while standing in front of a mirror, you press the underside tip of your tongue against the roof of your mouth.

As explained in Section 3.4.4.2, the tongue is a peculiar structure of the adult head, both developmentally and evolutionarily, because its muscles actually derive from the somites. To form the tongue, trunk mesodermal myocytes migrated around and beneath the pharynx and then rostrally to the oral region. Correspondingly, in humans these muscles are innervated by the **hypoglossal nerve (CN XII)** and by C1 (which runs with the hypoglossal nerve to innervate the geniohyoid muscle). The tongue includes the **root of the tongue** (posterior one-third), the **body of the tongue** (anterior two-thirds), and the **apex of the tongue**; these regions and the **dorsum of the tongue** are shown in Plate 3.40. The dorsum of the tongue includes the **foramen cecum** lying in the midline at the point of the **terminal sulcus (sulcus terminalis)**; this foramen marks the origin of the thyroglossal duct, formed by the caudal migration of the embryonic thyroid gland from an invagination of the tongue epithelium. Anterior to the terminal sulcus, the dorsum of the tongue is covered with **lingual papillae** (vallate, filiform, fungiform, and foliate) and is divided in the midline by the **median sulcus**. Posterior to the terminal sulcus lies the lingual tonsil. The **median glossoepiglottic fold** is a midline mucosal fold connecting the dorsum of the tongue to the **epiglottis** and lying medial to the lateral glossoepiglottic folds; the depressions between the median and lateral glossoepiglottic folds are the **epiglottic valleculae**.

There are four true extrinsic tongue muscles: genioglossus, styloglossus, geniohyoid, and hyoglossus (Plate 3.41; described in detail in Section 3.4.4.2). All are innervated by the **hypoglossal nerve (CN XII)** except the **geniohyoid**, which is innervated by the **C1 nerve** that runs along with the **hypoglossal nerve (CN XII)**. The intrinsic muscles of the tongue are the **vertical tongue muscles, transverse tongue muscles, superior longitudinal muscles**, and **inferior longitudinal muscles**, all of which are also innervated by the hypoglossal nerve.

The major vessels and nerves that supply the tongue include the **lingual artery**, which gives rise to the **deep lingual artery**; the **lingual vein**; the **lingual nerve**; and the hypoglossal nerve (Plate 3.41; described in detail in Section 3.3.2.1). The latter two structures pass between the hyoglossus muscle and the mylohyoid muscle, along with the **submandibular duct**. The lingual nerve gives rise to branches supplying the mucosa of the anterior two-thirds of the tongue with taste fibers and general sensation, as well as carrying postganglionic parasympathetic fibers from the **submandibular ganglion**.

3.5.12 Larynx

The larynx is a cartilaginous structure made up of the **thyroid cartilage**, **epiglottic cartilage**, **cricoid cartilage**, and **arytenoid cartilages** (Plates 3.43 and 3.44; described in detail in Section 3.2.3). The thyroid cartilage, formed of two **thyroid laminae**, **superior horns**, and **inferior horns** and connected to the hyoid bone superiorly by the **thyrohyoid membrane**, is the largest of these cartilages. Inferior to the thyroid cartilage and connected to it by the **cricothyroid joints** is the cricoid cartilage, composed of the **cricoid lamina** and **cricoid arch**. The thyroid and cricoid cartilages form a hollow tube and the epiglottic and arytenoid cartilages lie within. The epiglottic cartilage is suspended by its **stalk** from the thyroid laminae, whereas the arytenoid cartilages are attached to the cricoid cartilage by synovial joints and by the **vocal ligament** attached to the **vocal process** of the arytenoid cartilage.

The **laryngeal cavity** is shaped like an hourglass, composed of the **laryngeal vestibule** above the **vestibular folds** (**false vocal folds**), and the **infraglottic cavity** inferior to the vocal folds (which include the **vocal ligaments**) (Plate 3.44). The vestibular fold and the **vocal fold** (**true vocal fold**) are separated by a space called the **laryngeal ventricle**, which is quite variable in extent. The **glottis** includes the vocal folds and the space between them, which is designated the **rima glottidis**.

The five **intrinsic, or true, laryngeal muscles** (Plate 3.44; described in detail in Section 3.4) are the **posterior cricoarytenoid, lateral cricoarytenoid, arytenoid, thyroarytenoid**, and **vocalis**. The so-called **extrinsic laryngeal muscles** (described in Sections 3.5.1 and 3.5.8) are not true laryngeal muscles. All true (intrinsic) laryngeal muscles are innervated by the **inferior laryngeal nerve**, which is continuous with the **recurrent laryngeal nerve** and is connected to the **internal laryngeal**

nerve. Developmentally and evolutionarily, the **cricothyroid muscle** (Plate 3.21) is not a true laryngeal muscle either; it is instead a pharyngeal muscle derived from the inferior pharyngeal constrictor, being accordingly innervated by the **external laryngeal nerve** that also innervates the inferior constrictor.

3.5.13 Ear

BOX 3.24   DEVELOPMENT OF THE EAR: ONE ORGAN, THREE COMPARTMENTS

The **outer ear** consists of the **pinna** (**ear lobe, auricle**) and the **external ear canal**, which extends from the **external auditory meatus** to the **tympanic membrane** (**ear drum**) (Plate 3.45). In the embryo, the outer ear forms from several small swellings called **auricular hillocks** that develop along the boundary between branchial arches 1 and 2. The 2nd arch hillocks fuse together and with the dorsal 1st arch hillocks to form the **pinna**. The most ventral 1st arch hillock forms the **tragus**, which is a distinct and often separate cartilaginous flap located rostral to the ear canal (tragus is the Greek word for goat; in humans the tragus is often covered with hairy skin and the name is an allusion to a goatee). The sensory innervation of the external ear by branches of the **mandibular nerve** (CN V) and **facial nerve** (CN VII) nerves reflects its origin from branchial arches 1 and 2. All the muscles that move the pinna are facial expression muscles derived from 2nd branchial arch myoblasts that spread around the superficial aspects of the head and face, and thus are innervated by branches of the facial nerve (CN VII) (see Plate 3.27 and Section 3.4.2.1). As the head grows and expands outwardly, the deep part of the groove between the dorsal aspects of the 1st and 2nd branchial arches remains in a fixed position adjacent to the distal margin of the 1st pharyngeal (branchial) pouch; as a result, the external ear canals elongate. During much of gestation, the external ear canal is occluded by proliferating epithelial cells and their secretions; these dissolve before birth. We think of the outer ear and auditory canal as being on the dorsolateral surface of the head. But where are they at early stages in Plate 2.1? Ventrolateral! This apparent shift in location results from the tremendous ventral elongation of branchial arches 1 and 2.

The **middle ear** (Plate 3.45) is an air-filled epithelial-lined cavity that is continuous with the nasopharynx through the pharyngotympanic tube. Within this cavity lie three small bones, the **ear ossicles**. During development, the tubular part of the middle ear originates from the dorsal wing of the 1st pharyngeal (branchial) pouch; the cartilages derive from the **neural crest**, and the associated small muscles from 1st and 2nd arch-related mesoderm cells. The **middle ear ossicles** appear initially as three small aggregates of neural crest cells within the dorsal aspects of branchial arches 1 and 2, situated between the **otic vesicle**, the 1st pharyngeal (branchial) pouch, and the deep margin of the external ear canal (Plate 3.2). Mesenchymal cells around these cartilages degenerate, and the 1st pharyngeal pouch expands dorsally and laterally, and, together with some neural crest cells that transform to epithelium, completely surround the middle ear ossicles, and establish the **middle ear cavity**. The retained connection with the pharynx is the **auditory tube** (or **pharyngotympanic tube**, or **Eustachian tube**). The 1st pharyngeal pouch endoderm becomes closely apposed to surface ectoderm of the external ear canal, but the two layers remain separated by a layer of mesenchymal cells. This trio forms the definitive **tympanic membrane**. Myoblasts from 1st and 2nd branchial arch muscle primordia move into this region to form respectively the **tensor tympani** muscle, innervated by the **trigeminal nerve** (CN V), and the **stapedius muscle**, innervated by the **facial nerve** (CN VII). The ventral part of the middle ear cavity is the last to become enclosed with bone, which starts as a horseshoe-shaped ring beneath the skull and later forms a complete cup, the **tympanic bulla**. In animals that specialize in capturing low frequency sounds, the bulla becomes greatly enlarged.

The **inner ear** (Plate 3.45) is fully enclosed in the bony **petrous temporal bone** and has membranous, neural, and skeletal components. The entire fluid-filled and fluid-surrounded **membranous labyrinth**—including specialized **hair cells** and the neurons of ganglia of the **vestibulocochlear nerve** (CN VIII) that transmit auditory and vestibular signals to the brain—is derived from the **otic vesicle**, which forms initially as an **otic placode** (Plate 3.2) located beside the **hindbrain** and dorsal to the 2nd branchial arch. Soon after the otic vesicle forms, several dorsal and ventral outpocketings appear (Figure 3.8a). Three of these outpocketings emerge from the dorsal region. Each is shaped like half a

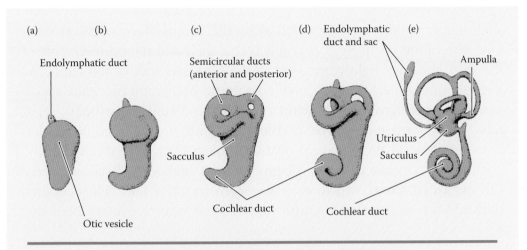

Figure 3.8 **(a–e) Morphogenesis of the membranous labyrinth.**

disc and is oriented in a different plane. In the center of each outpocketing, the epithelial sides fuse together and then degenerate, converting each disc to a macaroni-shaped tube called a **semicircular duct** (Figure 3.8). Specialized epithelial cells within enlargements (**ampullae**) at the base of each semicircular duct differentiate as mechano-receptive **hair cells**. Their function is to detect changes in the flow of fluid within each duct and thereby provide information on orientation and movements of the head. From the ventral part of the otic vesicle, a long slender evagination, the **cochlear duct** develops. This duct elongates and curls in a spiral, the diameter of the lumen becoming progressively smaller toward the end of the duct. From the medial surface of the otic vesicle, epithelial cells delaminate (break off) and form neuroblasts of the vestibular ganglion. Their axons project peripherally to hair cells in ampullae and the cochlear duct, and centrally into the brainstem. The central part of the original otic vesicle forms chambers called the **utricle** and **saccule**.

During morphogenesis of the inner ear, the membranous labyrinth is loosely encased in cartilage, which is able to remodel rapidly to accommodate the many changes in shape of the ducts (Figure 3.8). The space between the cartilaginous canals and membranous ducts is filled with **perilymph** and the ducts are filled with **endolymph**. Clusters of otic chondrocytes degenerate forming lacunae (lakes) that coalesce with and thereby enlarge the canal around the cochlear and **semicircular canals**. One band of chondrocytes is spared from degenerating and this forms a shelf that protrudes into the canal along the length of the cochlear duct. One side of the cochlear duct contacts this shelf and the opposite side of the duct becomes tightly apposed to the opposite wall of the canal. As a result, the formerly single cavity containing perilymph becomes subdivided into two cavities (the **scala tympani** and **scala vestibuli**), separated along the length of the cochlea with only a small communication between them at the distal tip of the cochlear duct. The site where the cochlear duct contacts the cartilaginous shelf marks the location where the **organ of Corti** will form. This organ includes hair cell receptors, derived from otic epithelium, and supporting connective tissue elements.

In the adult human, the **ear** includes the **internal ear** comprising the **vestibulocochlear organ** (the neurological part of the ear), the **middle ear** comprising the **tympanic cavity** and the **ear ossicles**, and the **external ear** comprising the **external acoustic meatus** (proximal two-thirds bone, distal one-third cartilaginous) and the **auricle** (Plate 3.45). The auricle includes the **auricular lobule** (part of auricular where there is no **auricular cartilage**), **antitragus**, **tragus**, **concha**, **antihelix**, and **helix**. The external and internal features of the temporal bone were described in Section 3.2.1.4 and shown in Plates 3.8a and c, 3.10 (including externally the **external acoustic meatus, mastoid process, stylomastoid foramen, jugular fossa, carotid canal,** and **bony portion of the pharyngotympanic tube**, and internally the **groove for the greater petrosal branch of the facial nerve, tegmen tympani,** and **internal acoustic meatus**). The tympanic cavity of the temporal bone is separated from the external acoustic meatus by the **tympanic membrane** and has the following borders: laterally, the tympanic membrane; medially, the **aditus** (opening posteriorly into the **mastoid air cells**), **promontory** (anterosuperior to the **round window**), and **oval window** containing the base of the **stapes**; anteriorly, the opening of the **pharyngotympanic tube**; posteriorly, the **tegmen tympani**; and inferiorly, the bony floor of the tympanic cavity.

The muscles that move the ear ossicles (**stapes, incus,** and **malleus**; described in Section 3.4) are muscles of the 1st (**tensor tympani**) and 2nd (**stapedius**) branchial arches. These muscles underwent dramatic changes during the origin of mammals, shifting from the region of the lower jaw to the region of the ear and becoming associated with hearing. The stapedius is therefore innervated by the facial nerve (CN VII) and, as its name indicates, is attached to the stapes. The tensor tympani is innervated by the trigeminal nerve (CN V) and is attached to the handle of the malleus, thus tensing the tympanic membrane as its name indicates.

Regarding the nerves of the ear region, the glossopharyngeal nerve (CN IX) innervates the mucosa of the tympanic cavity, forming a **tympanic plexus**. Its fibers then coalesce to form the **lesser petrosal nerve** that emerges through the tegman tympani just lateral to the **greater petrosal branch of the facial nerve**. The lesser petrosal nerve exits the cranial cavity through the foramen ovale and carries preganglionic parasympathetic fibers to the **otic ganglion**. The postganglionic fibers travel with the **auriculotemporal nerve** to reach the **parotid gland**, which it innervates.

The **facial nerve** (CN VII) follows a complicated course through the **temporal bone**, giving off several branches (Plate 3.16; described in detail in Section 3.3.1.7). It enters the **internal auditory meatus** and performs a hairpin turn behind the **cochlea**. At the anterior point of the turn, the **geniculate ganglion** is found. This is a sensory ganglion, whose axons carry information from the taste receptors of the anterior tongue and from the skin of the external auditory meatus. A branch of the facial nerve projects anteriorly from the geniculate ganglion, forming the greater petrosal branch of the facial nerve. These fibers emerge from the **tegmen tympani of the temporal bone** into the **middle cranial fossa**, and join the deep petrosal nerve to form the **nerve of the pterygoid canal (Vidian nerve)**. They are preganglionic fibers running to the pterygopalatine ganglion. The next branch is the **chorda tympani**, which passes between the **incus** and **malleus** near the **tympanic membrane** and then exits the temporal bone through the **petrotympanic fissure**, just posterior to the **mandibular fossa**, to join the **lingual nerve**. The third branch goes to the **stapedius muscle** in the middle ear. The rest of the facial nerve is entirely motor and exits the temporal bone through the **stylomastoid foramen**.

BOX 3.25 CONGENITAL HEARING LOSS

About 1 in 2000 infants is diagnosed annually with significant **hearing deficits**. Some deficits are associated with prematurity, ototoxic drugs, or infection (e.g., cytomegalovirus), but more than half are attributable to genetic factors. The majority affect the sensory apparatus—**hair cells** and sensory innervation. The most frequent mutations identified are in the *GJB1* gene, which encodes for a connexin26 protein. Inherited as an autosomal recessive trait, this mutation disrupts gap junctions between hair cells, leading to abnormal potassium flow and inability to respond to stimulation. Variable loss of melanocyte function in the skin and irises (causing a blue color) are associated with secondary loss of cochlear hair cells in humans (Waardenburg syndrome). Several different genes have been identified as causal factors, but the exact role that crest-derived melanocytes play during cochlear maturation and survival is not known.

Chapter 4

Upper Limb

The upper limb, also called the pectoral limb/appendage (or forelimb in most other tetrapod animals), is divided into the **shoulder**, **arm (brachium)**, **forearm (antebrachium)**, and **hand (manus)**. Details about each of these anatomical regions are given in the sections below.

BOX 4.1   BASIC CONCEPTS OF POSTCRANIAL MUSCULOSKELETAL DEVELOPMENT AND PRIMAXIAL VERSUS ABAXIAL MUSCLES

Throughout the **trunk**, cells located at the ventro-lateral margin of sclerotomes form sheets that expand laterally and then ventrally beneath the body wall. In the thoracic region, cells from each sclerotome aggregate to form the cartilaginous **ribs** and, at the upper limb level, some also form most of the **scapula**. The **costal (rib) cartilages** are initially fused with **vertebrae**, but later separate to establish moveable joints. Ossification of ribs begins mid-shaft and spreads proximally and, variably, distally (see also Section 6.1 for development and anatomy of the vertebrae and ribs). The sheets of **sclerotome**-derived connective tissue cells expand ventrally but do not reach the ventral midline. There, a separate population of lateral mesoderm-derived cells, which were brought to the ventral midline during closure of the body folds, forms the **sternum**. Initially, the sternum appears as paired, longitudinal cartilaginous condensations on either side of the ventral midline. These sternal bars subsequently fuse in the midline. As the cartilaginous **ribs** elongate ventrally, the more cranial 8 or 9 contact and transiently fuse with the sternum. The sternum begins to ossify between these sites of contact, giving it a segmental appearance (= **sternebrae**). Later, it fully ossifies. The distal cartilages of the more caudal ribs do not contact the sternum directly, but rather join with the adjacent, cranial rib. Ribs with distal ends unattached are called "**floating ribs**."

In contrast to the **shoulder girdle** (or **pectoral girdle**), the **hip bone** (**os coxae**) of the **pelvic girdle** receives no contribution from **somites**. *Rather, the hip bone, the **clavicle**, and all other skeletal elements of the limbs develop entirely from lateral somatic mesoderm of the body wall*. All skeletal muscles in the body develop from **paraxial mesoderm**—somites in the trunk and unsegmented mesoderm in the head.

Anatomists traditionally group muscles topographically. In the trunk, muscles are categorized as trunk **epaxial muscles** and **hypaxial muscles**, and as **appendicular muscles** based on their location and innervation. Epaxial muscles are those located *dorsal* (posterior in human bipedal anatomy) to the transverse processes of the vertebrae, are innervated mainly by **dorsal rami**, and mainly *extend* the back and move the neck. Hypaxial muscles are those associated with the body wall and vertebral column *ventral* (anterior in human bipedal anatomy) to the transverse processes of the vertebrae. They are innervated by **ventral rami** and *flex* the back (i.e., arch the back as a cat does when frightened). All the muscles of the thoracic and abdominal walls are also considered hypaxials; these include the **intercostal muscles** and **abdominal obliques**, including the **rectus abdominis**. Appendicular muscles that attach completely to limb skeletal structures (e.g., scapula, humerus, ulna, radius, and wrist/hand bones) are considered **intrinsic limb muscles**; those that join the limb to axial skeletal structures (e.g., sternum or vertebrae) are **extrinsic limb muscles**, mainly derived from myotomes.

Developmental biologists, who study where muscle progenitors originate and how they disperse, use a different scheme of trunk muscle categorization. There are two main categories: **primaxial muscles**, which correspond to epaxial and some appendicular muscles (mainly extrinsic limb muscles), and **abaxial muscles**, which correspond to hypaxial and other appendicular muscles (mainly intrinsic limb muscles). One of the main differences between the two terminologies is that the terms "epaxial" and "hypaxial" refer to *innervation*, epaxial muscles being innervated mainly by **dorsal rami**, and hypaxial muscles by **ventral rami**. In contrast, the terms "primaxial" and "abaxial" refer to different myotome subpopulations, following recent developmental studies. On the basis of mapping data, Ann Burke (from Wesleyan University) and colleagues identified two populations of myogenic progenitors within the outer, myotome part of each somite. The **primaxial myotome** has a population of myoblasts that, as its name indicates, arises close to the body axis, adjacent to the neural tube from which muscle-inducing signals are delivered. These myoblasts have limited migratory potential, and thus form the epaxial and proximal hypaxial muscles that generally remain close to the vertebrae. While they are not migratory, these primaxial myocytes will move together with sclerotome cells that expand part of the way around the body wall (which give rise to the so-called epaxial muscles of the back, to the so-called proximal hypaxial muscles of body wall [e.g., intercostals], to neck muscles such as the scalenes, and the so-called extrinsic limb muscles such as the rhomboids, levator scapulae, and latissimus dorsi). Myoblasts of the **abaxial myotome** arise along the ventro-lateral margin of the myotome, are influenced primarily by signals from the adjacent lateral plate mesoderm and ectoderm, and migrate independently of other somitic cells to populate the more ventral part of the body wall and the limbs. As explained in Section 3.4.4, the only populations in the head and neck that are similar to the long-distance migrating limb myoblasts are those that emigrate from anterior somites to form the infrahyoid and tongue muscles (i.e., the **hypobranchial muscles**).

4.1 Upper Limb Skeletal System

BOX 4.2 LIMB DEVELOPMENT

Tetrapodism (i.e., having four appendages, two pectoral ones and two pelvic ones) is present in most jawed vertebrates, and elaboration of **appendages** has played a pivotal role in the ability of species to adapt to niches as diverse as land, sea, and air. The origin and evolution of vertebrate appendages has been primarily accomplished by alterations in the number of distal elements and tremendous variations in the relative lengths of each limb bone (Plate 4.1b). However, limb reduction or loss is also a common theme in some groups of vertebrates, including snakes and cetaceans (whales and dolphins). In a further example of **developmental constraints** and **phylogenetic constraints** reflecting evolutionary history, embryos of animals belonging to these groups still initiate the development of four limb buds. Among mammals, there are substantial differences in the posture and proportions of different parts of the limbs, with the most obvious variations seen in the most distal elements, the digits.

Despite this diversity of adult phenotypes, early limb development is remarkably similar across tetrapods. Tetrapod upper and lower limbs consist of a single proximal element (**stylopod**: **humerus** or **femur**) attached proximally to one or several girdle elements (**scapula** or **hip bone**), a paired set of middle elements forming the **zeugopod** (**radius/ulna** or **tibia/fibula**), and a variable number of distal nodular and radial elements, which collectively are called the **autopod** (**hand** or **foot**).

Limb development begins after the **neural tube** (Figure 3.4), **somites**, intermediate **mesoderm**, and both somatic and splanchnic layers of **lateral plate mesoderm** have formed. The first overt sign of limb development is a focal increase in cell proliferation in lateral **somatic mesoderm** at four sites along the body wall adjacent to somites. This proliferation creates epithelial-covered swellings, called **limb buds** (Plate 4.2), which project outward from the body wall. These buds are covered by surface **ectoderm** and arise at sites corresponding to the axial levels for their innervation. As the lateral body folds develop, these limb buds shift their orientation to a more lateral position. Each limb bud elongates proximally-to-distally, and becomes flattened dorso-ventrally. Along the distal tip of each limb bud, surface ectoderm forms a longitudinal thickening called the **apical ectodermal ridge** that provides growth factors essential to maintain the distal growth of the early limb bud. Mesenchymal cells located beneath this ridge continue to proliferate rapidly, while further from the ridge cell division rates diminish and chondrogenesis is initiated. This distal region of rapidly dividing mesenchymal cells is called the **proliferative zone**.

While all intrinsic appendicular connective tissues are derived from limb buds, girdle elements may derive from the same (e.g., hip bone) or multiple *(scapula, clavicle) sources, including the somites. Except for the **clavicle**, all limb skeletal tissues form initially as cartilage and ossify secondarily.* Limb development progresses from proximal to distal. The humerus (or femur) differentiates first; then a pair of antebrachials (the radius and ulna, or the tibia and fibula) appears. Next to form are the **carpals/tarsals (wrist/ankle)**, then **metacarpals/metatarsals**, and then the **phalanges**. Each element arises as a distal extension of the more proximal, previously formed cartilage condensations (Plate 4.2). The basic configuration for distal structures in modern tetrapods is to form five **digits (pentadactyly)**, but in many animals, some of these either fail to form or show arrested growth early in development (**nonpentadactyly**, one of the most common limb **congenital malformations** in humans). An important component of limb morphogenesis is **programmed cell death**. Cells located at sites where joints form and also between developing digits degenerate at the end of their differentiation. This distal programmed cell death is reduced in bat wings and duck feet, resulting in the retention of webbing between the digits.

The dorso-ventral (postero-anterior in human anatomy), antero-posterior (radio-ulnar or tibio-fibular in human anatomy), and proximo-distal axes of the limb are established in the early limb bud stages. The limb bud is in fact asymmetric in both the antero-posterior and dorso-ventral axes. The apical ectodermal ridge extends further along the posterior side, and the dorsal surface shows greater convexity. Asymmetries in the dorso-ventral axis become more evident when structures unique to the integument (e.g., **nails** or **claws**) begin to differentiate. However, internally, there are many dorso-ventral asymmetries in the attachments of tendons and remodeling of joints.

Limb formation is initiated at the 6–7 mm stage in most mammals, corresponding in humans to week 5 (Plate 2.1). By the time an embryo reaches 25 mm, the limbs are about 6 mm long and most internal structures are established, including the neuromusculoskeletal elements and joints. Next comes a prolonged period of remodeling and growth, driven by skeletal remodeling and dependent on proper functioning of all systems. As noted above, limb buds initially grow dorsolaterally, perpendicular to the body wall of the early embryo, but as the limbs elongate, they progressively shift their orientation both with respect to the body and within the limb. Limb buds are invaded by skeletal

myoblasts from somites, **angioblasts** from paraxial mesoderm, and **axons** from the **spinal cord** and **sensory ganglia**. The lower limbs are fixed in position by their articulation with the pelvis (os coxa), which is firmly attached to lumbosacral vertebrae. In contrast, the weight-bearing scapula of the forelimb girdle is not directly articulated with the **vertebral column**, being instead indirectly articulated to it through the clavicle.

The adult human **upper limb skeleton** is formed by the **pectoral girdle** (or **shoulder girdle**) and the bones and cartilages of the **arm**, **forearm**, **wrist**, and **hand**. Most of these elements are derived from endochondral ossification, as noted above. The main exceptions are the **clavicle** and **scapula**, which together form the **pectoral girdle**.

4.1.1 Pectoral Girdle

The clavicle or **collarbone**, is a long bone that connects the scapula to the **sternum** via the acromioclavicular and sternoclavicular joints (Plates 4.5, 4.7, and 4.8). The **acromioclavicular joint** is a plane synovial joint between the acromion of the scapula and the lateral end of the clavicle, and is associated with the **coracoclavicular ligament**—formed by the **conoid ligament** and the **trapezoid ligament**—and the **coracoacromial ligament** (Plate 4.8b). Note that the clinical condition known as **shoulder separation** refers to a separation between the acromion and the clavicle, that is *within* the shoulder girdle and not to a separation *between* this girdle (and namely the scapula) and the humerus (arm), which is called **shoulder dislocation**. The medial end of the clavicle articulates with the **clavicular notch of the manubrium** and the adjacent part of the **1st costal cartilage**, forming the **sternoclavicular joint**. This joint is stabilized by the **anterior sternoclavicular ligament** and the **costoclavicular ligament**, and its **articular disc** prevents medial displacement of the clavicle (Plate 4.8a). The saddle shape of the joint surfaces, combined with the articular disc, allows the sternoclavicular joint to function almost as freely as a ball-and-socket joint. As the scapula is directly connected to the clavicle, the sternoclavicular joint is crucial for the protraction (anterior displacement), retraction (posterior displacement), depression, elevation, medial rotation—inferior angle going medially—and lateral rotation—inferior angle going laterally—of the scapula.

In addition to these six movements, the scapula can also be adducted (i.e., moved medially, toward the midline; in the human body this movement is normally coupled with a retraction of the scapula) by the **rhomboid major** and **rhomboid minor**, and abducted (returned to its original position) mainly by passive movement.

The clavicle is the first bone to begin ossification in the embryo, during the fifth and sixth weeks of gestation, and paradoxically one of the last bones to finish ossification, at about 21–25 years of age. The **lateral end of the clavicle** is formed by **intramembranous ossification**, while part of **shaft of the clavicle** and the **medial end of the clavicle** are formed by **endochondral ossification**.

The larger portion of the scapula (Plate 4.7a and b) is formed by intramembranous ossification, while the outer parts are mainly formed by endochondral ossification. Along the superolateral portion of the scapula are the **acromion, suprascapular notch** (bridged by the superior transverse scapular ligament), and **supraspinous fossa**. The **scapular spine** extends from the acromion to divide the **supraspinous fossa** and **infraspinous fossa** (located superior and inferior to the spine, respectively). On the lateral aspect is the **glenoid fossa** (or **glenoid cavity**, deepened by the cartilaginous **glenoid labrum**), which articulates with the humerus. The **supraglenoid tubercle** lies just superior to this cavity, while the **infraglenoid tubercle** lies inferior. Inferior to the **glenoid cavity** is the **lateral border of the scapula**, which, when followed medially, becomes the **inferior angle of the scapula**. The **medial (vertebral) border of the scapula** will become the **superior angle of the scapula** when followed superiorly. The **coracoid process** lies inferior to the **acromion** and just medial to the **glenoid cavity** (this process is the remnant of a bone present in our fish ancestors, the coracoid bone).

4.1.2 Shoulder Joint and Arm

The proximal portion of the **humerus** includes several gross features: **head of the humerus, anatomical neck of the humerus, surgical neck of the humerus**; processes for muscle attachment such as the **greater tubercle of the humerus, lesser tubercle of the humerus**, and **deltoid tuberosity**; and grooves for passage of tendons and vessels such as the **humeral intertubercular sulcus (bicipital groove)**, and **radial groove** (Plate 4.7). As their names indicate, the anatomical neck is the one that can be more easily seen in gross observation of dry bones, surrounding the articular surface, while the surgical neck is the one most at risk of fracture. The articulation between the head of the humerus and the glenoid cavity of the scapula is called the **glenohumeral joint** or **shoulder joint**. The **glenohumeral ligaments** connecting the humerus and scapula strengthen the anterior wall of the **capsule of the shoulder joint**. The shoulder joint allows the humerus to be highly mobile: It can be flexed, extended, adducted, abducted, medially (internally) rotated, and externally (laterally) rotated (see Box 4.3).

BOX 4.3 EVOLUTION AND PATHOLOGY: ARBOREAL PRIMATE ANCESTORS, SUPRASPINATUS TENDINITIS, AND PAINFUL ARC

The extreme mobility of the humerus, particularly the ability to abduct it up to 180° or more from the anatomical position, is a relic from our **arboreal primate ancestors**. This ability is associated with anatomical features such as having the **supraspinatus** lying on the superior—rather than the posterior (dorsal)—side of the scapula and shoulder joint. Other mammals such as dogs, horses, and rats obviously cannot perform this movement. Although humans maintain this ability, it is no longer essential for our survival as it was for our arboreal primate ancestors and is for most living arboreal primates (Figure 4.1). As often happens in evolution, some advantages—for arboreal primates, brought by derived features such as having a superiorly displaced supraspinatus—carry new disadvantages—in this specific case, for humans. This balance is called an **evolutionary trade-off**. Because the supraspinatus—unlike to the other "**rotator cuff**" muscles—lies on the superior region of the scapula and shoulder joint where many other structures also lie (for instance, cartilages, bursae, ligaments), its tendon is often pressed against the **acromion** as the humerus is abducted. Friction between the supraspinatus tendon and the acromion is usually reduced by a fluid filled sac lying between these two structures: the **subacromial bursa**. However, sometimes wear and tear results in **supraspinatus tendinitis**, associated with inflammation of the bursa: **subacromial bursitis**. Partial tears and tendinitis of the supraspinatus tendon causes the clinical condition known as **painful arc**, often seen in athletes such as tennis or baseball players.

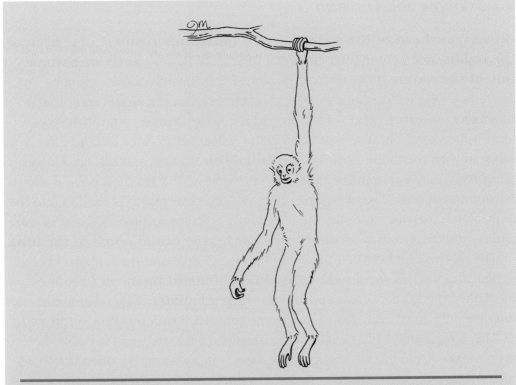

Figure 4.1 **Arboreal (tree-living) primate (gibbon) demonstrating the 180° abduction of the shoulder. While modern humans share this ability, the evolutionary trade-off is predisposition to syndromes such as "painful arc," and general risk of injury to the "rotator cuff" of the shoulder.**

The distal portion of the humerus includes the **medial epicondyle, medial supracondylar ridge, lateral supracondylar ridge,** and **lateral epicondyle,** all of which serve as muscle attachment sites. It also includes surfaces that articulate with the radius and ulna: the **coronoid fossa** to receive the **coronoid process** of the ulna, the **capitulum** that forms a gliding joint with the head of the radius, the **trochlea** that forms a hinge joint with the **trochlear notch** of the ulna, and the **olecranon fossa** that receives the **olecranon process of the ulna** (Plate 4.12a–c). An easy way to orient the distal portion of an isolated humerus is to check whether you are looking at the small coronoid fossa or the large olecranon fossa, meaning that it is an anterior or posterior view, respectively. Then, look for the medial epicondyle—which is far more pronounced than the lateral epicondyle—to identify the medial side.

4.1.3 Elbow and Forearm

Between the **head of the radius** and the **radial tuberosity** lies the **neck of the radius**, and posterior to the radial tuberosity lies the **anterior oblique line of the radius** (Plate 4.12a–c), which serves as a muscle attachment. There is a logic to the names of the characteristics of the radius and ulna. A tuberosity is a prominence of a bone, so the bone's name comes before the term "tuberosity," because that bone is the *active player*—for instance, the tuberosity on the radius is called the **radial tuberosity**. A notch on a bone often *receives* the prominences of *another bone*, so the notched bone is the *passive player* now. Therefore, the name of the active player is used before the term "notch," so, for example, the ulna has a notch to receive the head of the radius, and this notch is accordingly designated the **radial notch of the ulna**.

The elbow joint between the arm bone (humerus) and the forearm bones (radius and ulna) is stabilized by the **ulnar collateral ligament** (running medially, from humerus to ulna) and the **radial collateral ligament** (running laterally, from humerus to radius and annular ligament) (Plate 4.12d and e). This joint allows only flexion and extension of the forearm. Because there is no posterior ligament, backward dislocation is most common, often associated with fracture of the coronoid process. Continuous with the elbow joint, there is also a **proximal radio-ulnar joint** between the radius and ulna for the pronation and supination of the forearm. As a whole, the radio-ulnar joint includes this proximal part in which the radial notch of the ulna receives the head of the radius, which is encircled by the **annular ligament**, and a distal part in which the **ulnar notch of the radius** receives the **head of the ulna**. In the distal part of the joint, the ulna is also connected to the radius by an **articular disc**. The radius is free to rotate within the annular ligament, but due to the intermediate part of the radio-ulnar joint—the **interosseous membrane**—and the articular disc, the radius and ulna stay together during this rotation. On the lateral side of the distal radius lies the **styloid process**, a site of muscle attachment. On the medial side of the distal ulna lies a similar, but smaller, styloid process. Note that the heads of the humerus and radius lie on the *proximal* region of these bones, whereas the heads of the ulna and of hand bones such as the metacarpals and phalanges lie on the *distal* region of these bones.

4.1.4 Wrist and Hand

The **wrist joint (radiocarpal joint)** is the articulation between the proximal **carpal bones (scaphoid, lunate,** and **triquetrum)** and the distal end of the **radius**, and is reinforced by the **radiocarpal ligaments** (Plate 4.16a). This joint—which does not include the ulna—allows flexion, extension, adduction, abduction, and circumduction (circular movement) of the hand. The **tubercle of the scaphoid**, **hook of the hamate**, and **tubercle of the trapezium** are distinctive features of the carpal bones. In addition to these three bones and the lunate and triquetrum mentioned above, the carpal region includes the **trapezoid**, **capitate**, and **pisiform** bones. The pisiform is thought to differ from the other carpal bones in that it forms by **sesamoid ossification** (ossification of a ligament or tendon), probably the tendon of the flexor carpi ulnaris muscle. The **flexor retinaculum**, a fibrous band of connective tissue, extends from the pisiform and the hook of the hamate over the tubercles of the scaphoid and trapezium, forming the **carpal tunnel**.

The **midcarpal joint** is a joint between the proximal and distal carpals, which contributes to abduction and flexion, and particularly to adduction and extension, of the hand. The **carpometacarpal joint** between the distal carpals and the **metacarpals** does not contribute significantly to movements of the digits, except those of the **thumb** (digit 1), which is much more mobile than the other digits, and to a lesser extent digit 5. It is this thumb joint that allows the palmar surfaces of digits 1 and 5 to meet each other in full **opposition** (Figure 4.2). This ability is only present in humans, and reflects the crucial role played by **thumb movements** (for instance, for tool manufacture and use) in the evolutionary history of the human lineage.

As its name indicates, the **metacarpophalangeal joint** is a joint between the **metacarpals** and the **phalanges of the hand**. This joint allows flexion, extension, abduction, and adduction of the proximal phalanges and thus of the digits as a whole. Lastly, between the phalanges are **interphalangeal joints** (only one in the thumb, but two—proximal and distal—in the other digits). These joints only allow flexion and extension of the middle and distal phalanges.

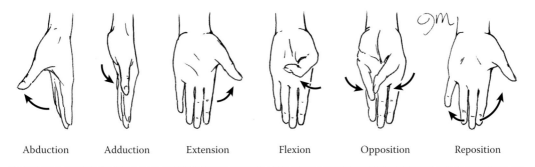

| Abduction | Adduction | Extension | Flexion | Opposition | Reposition |

Figure 4.2 Movements of the thumb.

Students tend to confuse the movements of abduction and adduction of the fingers, and particularly of **abduction, adduction, flexion, and extension of the thumb** (Figure 4.2). However, it is easy to learn these movements when you know the functional axis of the human hand. Because our arms and forearms are so mobile, it would be difficult to define the movements of abduction/adduction of the digits based on the midline of the body. Therefore, these movements are defined relative to a more logical axis: the axis of the hand, that is, digit 3 (middle finger) (Plate 4.16b and c). This digit can be abducted to both the ulnar and radial sides, that is, moving away from its anatomical position. Obviously, it cannot be adducted from this position, because any movement (either radial or ulnar) brings it away from the axis of the hand. As noted above, the thumb is a highly mobile digit in humans, and one of its particularities is that in the anatomical position, it is rotated 90° medially with respect to the other digits. Therefore, its dorsal (extensor) side is oriented laterally (radially), and not posteriorly like those of the other digits. Keeping this in mind, all the thumb movements make sense: Moving the thumb toward the side of its own nail (i.e., laterally/radially) is extension, moving toward the side of its pad (i.e., medially/to ulnar side) is flexion, and moving toward (i.e., posteriorly) and away (i.e., anteriorly) from digit 3—the axis of the hand—in a plane parallel to the nail, is adduction and abduction, respectively.

BOX 4.4   SOME BASIC CONCEPTS OF LIMB DEFECTS

With respect to underlying mechanisms of three-dimensional organization and tissue specification, limbs are the most thoroughly studied musculoskeletal structures. Many abnormalities have been produced experimentally, and some mimic those seen in patients. The following criteria are commonly used to assess **limb defects**:

1. *Primary versus secondary.* Secondary defects are those in which limb development is affected by lesions in other parts of the body. For example, loss of innervation results in lack of movement that causes developing joints to fuse (= **ankylosis**), usually in an abnormal posture (= **arthrogryposis**).
2. *Morphogenesis versus growth.* Spatial organization of the limb and all its tissues occurs during limb bud outgrowth and initial changes in posture. Defects in spatial organization during these stages are

considered *morphogenetic*. In contrast, a skeletal structure that forms but subsequently fails to grow properly may appear reduced or absent at term (birth), in which case the lesion affected its growth, not its initial formation. This distinction is important, because the interactions and most of the genes that promote early **limb morphogenesis** are different from those that promote limb growth and ossification.

3. *Limb axis affected.* A primary lesion can affect proximo-distal, dorso-ventral (postero-anterior in human anatomy), or antero-posterior (radio-ulnar or tibio-fibular in human anatomy) patterning. Many lesions that affect one of these axes will have secondary consequences on the others.

4. *Duplication versus reduction.* The presence of extra digits is **polydactyly**. Fusion of digits is **syndactyly**, and loss of digits is **ectrodactyly**. **Brachydactyly** is shortening of a digit, which can occur due to reduced growth or loss of one or several segments. Defects compromising other parts of a limb are described using the suffix *-melia* with these same prefixes (e.g., **amelia** = absence of entire limb; **ectromelia** = loss of part of limb; and **brachymelia** = shortened limb).

5. *Isolated versus syndromic.* Many limb defects are syndromic, and often the other abnormalities are of greater severity. In most cases, this outcome reflects pleiotropic functions of the affected gene(s) (i.e., a gene has many different functions) or shared sensitivity to a teratogen.

The early **limb bud** is not a miniature of the adult; rather, as noted above, the more proximal region-specific elements such as the humerus are established first, then the more distal structures, and finally the digits, whose progenitors emerge from within the mesenchyme immediately beneath the distal margin of the limb bud (Plate 4.2). It is not surprising, therefore, that the majority of developmental limb defects preferentially affect the later-forming, distal structures. The cell interactions necessary to establish and maintain these proximo-distal asymmetries have been known since the mid-twentieth century. They were revealed by manipulations of chick wing and leg buds. For instance, removal of the **apical ectodermal ridge** (AER) stops all additional distal outgrowth, and underlying mesenchymal cells degenerate. In contrast, the remaining, more proximal mesenchymal cells continue to differentiate, forming the skeletal tissues normally. These data indicate that a *important role of the AER is to maintain mitotic activity and/or survival in the underlying mesenchyme.* Many factors are involved in this signaling, key among which are members of the FGF (fibroblast growth factor) family.

If the epithelial covering of the limb bud is removed, rotated 90° around the **dorso-ventral limb axis** (postero-anterior axis in human anatomy) and placed over the exposed mesenchymal core, the limb develops with normal proximo-distal and antero-posterior axes (radio-ulnar or tibio-fibular in human anatomy), but with an inverted dorso-ventral axis. Not only are the skeletal structures and joints reversed, but so are the muscles. The latter outcome is surprising, because the myoblasts that form limb muscles arise outside of the limbs, in abaxial myotomes, and migrate into developing limb buds, as explained above. Clearly, then, the development of musculoskeletal relationships is a local process, not one decided within muscle progenitors. Defects of the dorso-ventral axis are much less common than those of other axes. One example of dorso-ventral axis defects is the **nail-patella syndrome**, an autosomal dominant condition in which the patellae are reduced in size or (less frequently) missing, and patients' fingernails are very narrow; these are variably accompanied by **elbow dysplasia**, **glaucoma**, and **kidney problems**. In most patients, mutations in a transcription factor called *LMX1B* have been found.

Similar experiments in chickens in which the antero-posterior axis was reversed initially gave inconsistent results. The distal part of the limb bud, including both epithelial and mesenchymal components, was rotated so as to reverse the antero-posterior axis (radio-ulnar or tibio-fibular in human anatomy). It was discovered that small differences in the amount of posterior (ulnar/fibular side) limb bud mesenchyme included in the graft made a profound difference to the outcome of the experiment. This region of posterior mesenchyme, which is located adjacent to the caudal margin of the AER, is called the **zone of polarizing activity** (ZPA). Transplantation of this mesenchymal tissue into the anterior (radial/tibial in human anatomy) part of a limb bud promotes formation of a new (or anteriorly elongated) AER, and thereby increases the cell division around the graft. This new AER results in the formation of a second limb, whose posterior (i.e., ulnar/fibular in human anatomy) margin is always close to the implant. This outcome contrasts with "typical" preaxial polydactyly (Figure 4.6), in which the extra digits have normal, not reversed asymmetry.

The ZPA acts in a concentration-dependent manner. Grafting a large piece of tissue, for instance, can create a digit pattern 4-3-2-2-3-4 (i.e., six digits, the first of which has four phalanges, etc.). Grafting a half-sized piece instead can produce 3-2-2-3-4, and implanting only a few hundred cells can result in a 2-2-3-4 pattern.

The signaling molecule released by this polarizing mesenchyme is **sonic hedgehog**, the same molecule produced by the notochord that initiates the formation of the sclerotome and establishes a ventro-dorsal gradient within the spinal cord. Grafting a bead soaked in sonic hedgehog to the anterior (radial/tibial in human anatomy) margin of the limb bud produces results identical to the ZPA grafts: extra digits, often with reversed antero-posterior orientation. This posteriorly (ulnar/fibular) restricted expression of sonic hedgehog is maintained by patterns of ***Hox* gene** expression. Early in vertebrae evolution, several members of the original eight-member gene family were duplicated and translocated 5′ (downstream) to the original family. Gene sequencing identified these new genes as *Hox* 9, 10, 11, 12, and 13. They are expressed only in limbs, and like other members of the *Hox* gene family they are activated in numerical order along the antero-posterior axis (see Box 6.1). *Hox* genes in groups 10 through 13 are activators of sonic hedgehog expression. Similar to their actions in body axis specification, the expression of each new *Hox* gene group occurs at a more posterior (ulnar/fibular) location, and is added to those activated earlier. As a result, the total number of Hox proteins increases along an antero-posterior (radio-ulnar or tibio-fibular in human anatomy) axis. In the limb, this ensures that the amount of sonic hedgehog produced will also increase posteriorly.

The link between *Hox* genes and sonic hedgehog expression came recently with the discovery of a noncoding gene sequence called ZRS that is responsive to Hox protein binding and is located upstream from the sonic hedgehog gene. Mutations at this site have been found in humans with inherited **polydactyly**. These mutations create a change in one amino acid, and as a result allow low levels of sonic hedgehog expression in the anterior (radial/tibial) mesenchyme of the early limb bud. This expression is sufficient to initiate the formation of extra digits, but is not enough to repolarize the expanded anterior (radial/tibial) part of the limb. As expected, variations along the antero-posterior axis (radio-ulnar and tibio-fibular in human anatomy) are common among mammals, including humans, being most evident in the number of digits. In addition to the number of digits present, there are also many differences in the anatomy of preaxial versus, postaxial digits, most evident in such evolutionary specializations as our peculiar **opposable thumb**. The most common malformations of the hand/foot involve the fusion of adjacent digits (**syndactyly**) or the presence of extra digits (**polydactyly**).

Polydactyly is uncommon in humans, but can be inherited and often in these cases is part of a syndrome (for instance, trisomy 18; Figure 4.6). **Pallister–Hall syndrome** is an autosomal dominant condition in which the extra digit is postaxial. These patients also have hypothalamic lesions. In **Greig cephalopolysyndactyly syndrome**, the duplicated digit is located preaxially. This autosomal dominant condition is highly variable in its other manifestations, but usually includes hypertelorism and an enlarged head. Both these syndromes affect the sonic hedgehog response pathway by preventing the cleavage and thus the positive effects of a transcription factor called **Gli3**. Normally, in the presence of elevated activation by sonic hedgehog, Gli3 is converted from a repressing to an activating factor. Greig cephalopolysyndactyly is a full null mutation of Gli3, which, therefore, reduces the available amount of this transcription factor by half (the remainder coming from the other chromosome). Pallister–Hall has a partially shortened and thus functionally reduced Gli3, which functions as a repressor but cannot bind as an activator.

4.2 Upper Limb Neurovascular System

4.2.1 Upper Limb Nerves

A brief summary of some basic concepts of the development of the nervous system, including references to limb innervation and to experiments using model organisms such as chickens, was given in Section 3.3.1 and Box 3.11. Notes on the development and anatomy of the peripheral nervous system were also given in Section 3.3.1, in Box 3.12.

4.2.1.1 Brachial Plexus

Knowing the organization of the **brachial plexus** (Figure 4.3) is critical for understanding the motor and cutaneous innervation of the upper limb. Many students memorize the brachial plexus using mnemonic devices, but most aspects of its organization can be more easily understood in the context of the evolutionary and developmental constraints of the body. For example, if an engineer without previous anatomical knowledge looked at the human body, some aspects of it would seem to make no sense. For example, nerves from **ventral rami** use peculiar—and often risky (subject to injury) "tricks" to get to the back of the body to innervate **posterior pectoral shoulder muscles**. To an engineer, it would seem to be much easier for nerves from **dorsal rami** to innervate those posterior (dorsal) muscles. However, the configuration of these nerves in the adult human makes complete sense if you take into account our developmental and evolutionary history, as explained in Section 4.3.

The four "tricks" that allow nerves from the ventral rami to innervate posterior (dorsal) muscles, shown in Plate 4.7b, are

1. *Axillary nerve (lateral trick)*: To reach the posterior shoulder, the axillary nerve branches from the distal bifurcation of the posterior cord to pass inside the *laterally* located **quadrangular space** of the back.
2. *Dorsal scapular nerve (medial trick)*: To reach the medial side of the back, the dorsal scapular nerve branches early (i.e., *medially*) from the brachial plexus, running posteriorly (dorsally, as its name indicates) to pass medially to the scapula.
3. *Suprascapular nerve (superior trick)*: To reach the posterior aspect of the pectoral girdle, this nerve, as its name indicates, passes through the **suprascapular notch of the scapula**, which is located on the *superior* aspect of the scapula the scapula.
4. *Subscapular nerves (inferior)*: To reach posterior muscles, these nerves run inferiorly, passing anterior (ventral) to the scapula, some of them reaching a region that lies *inferior* to the scapula (i.e., the thoracodorsal nerve, or middle subscapular nerve, that innervates the latissimus dorsi).

Therefore, here we lead the students to understand and be able to build—mostly logically—the brachial plexus, step by step, from its five trunks to all the major nerves that directly branch from it.

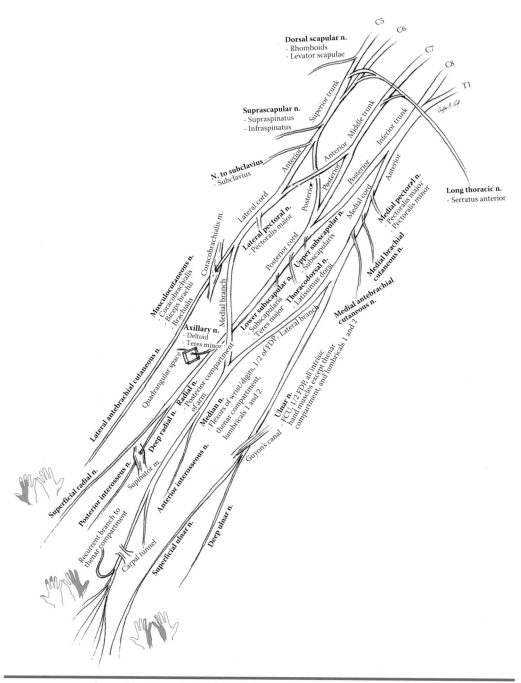

Figure 4.3 Diagram of the brachial plexus showing divisions and courses of each nerve and their target tissues, schematic view. (Modified from learning tool developed by the VCU School of Medicine, Class of 2009.)

BOX 4.5   EASY WAYS TO REMEMBER: ROOTS, TRUNKS, DIVISIONS, AND CORDS OF BRACHIAL PLEXUS

The **roots of the brachial plexus** are the ventral rami of C5, C6, C7, C8, and T1 (Figure 4.3). An easy way to memorize the configuration of the brachial plexus as a whole is to begin with the central root, C7. C7 behaves in a peculiar way, somewhat the "ugly duckling" of the brachial plexus. For instance, this root mainly gives rise to the posterior cord and radial nerve, which lie posterior to the other parts of the brachial plexus. Therefore, it is not surprising that C7 is the only root that will not meet another root to form the **trunks of the brachial plexus**. Instead, C7 forms its own **middle trunk**, which lies between the **superior trunk** formed by C5 and C6 and the **inferior trunk** formed by C8 and T1.

Next, each trunk divides into two divisions, anterior and posterior, for a total of six **divisions of the branchial plexus**. Logically, the division that is mainly a continuation of the middle trunk (from the "ugly duckling" C7) is a posterior division. Accordingly, the divisions of the other two trunks that meet this division (branching inferiorly from the superior trunk and superiorly from the inferior trunk) are posterior divisions as well, while the divisions that are mainly continuations of the superior and inferior trunks are the anterior divisions. As might be expected, the three posterior divisions form the **posterior cord**, while the anterior divisions that derive from the superior and inferior trunks will be parts of the **lateral cord** and **medial cord**, respectively. Since three of the six divisions of the brachial plexus form the posterior cord, one of the other two cords will have two divisions, while the other has only one. The question is, which cord receives an anterior division, from the middle trunk? This is one of the few exceptions that students need to memorize, because it cannot be easily explained by using simple evolutionary/developmental explanations: The anterior division derived from the middle trunk goes *superiorly* to meet with the anterior division derived from the superior trunk, forming the lateral cord.

Lastly (more distally), the cords bifurcate. The lateral cord gives rise to the **musculocutaneous nerve**, which as expected is the most lateral of the three major deep nerves of the arm. Similarly, the medial cord gives rise to the most medial of these three nerves, the **ulnar nerve**. The nerve that lies between the ulnar and musculocutaneous nerves at the level of the arm is the **median nerve**. The **lateral branch/root of the median**

nerve thus comes from the lateral cord, while the **medial branch/root of the median nerve** comes from the medial cord. As expected, the radial nerve (which lies mainly in the posterior region of the arm and hand) is a branch of the posterior cord. The **axillary nerve** is the other branch of the posterior cord; its origin from the posterior cord is also logical, because *it performs the first of the four "tricks" (i.e., the "lateral trick") listed above, passing through the quadrangular space to innervate two of the most lateral muscles of the posterior pectoral girdle,* the **deltoid** and **teres minor** (shown in Figure 4.3 and explained in Section 4.3).

The roots, trunks, divisions, and cords of the brachial plexus, as well as their distal bifurcations, are explained in the box above. So, what about the nerves that branch directly from the more proximal portion of the brachial plexus? Two nerves branch from this plexus at the level of the roots: the dorsal scapular nerve and the long thoracic nerve (Figure 4.3). The **dorsal scapular nerve** is named logically; as will be explained in Section 4.3, this nerve supplies posterior muscles (*dorsal* muscles, in comparative anatomy) that attach to the *scapula*; namely, the **levator scapulae, rhomboid major**, and **rhomboid minor**. *Because these three muscles lie on the medial side of the back, the dorsal scapular nerve thus performing the second of the four "tricks" (i.e., the "medial trick") listed above to reach them by branching early and passing medially to the scapula* (Plate 4.7b). The other nerve to branch from the roots is the long thoracic nerve—its name also makes sense because it is a *long* nerve that has to run all the way down (inferiorly) to the inferior *thoracic* region, to innervate the **serratus anterior**.

Two nerves branch off at the level of the trunks (Figure 4.3), both from the superior trunk, one running more anteriorly and the other more posteriorly. The more anterior one is the **nerve to subclavius**, which as its name indicates innervates the **subclavius muscle**. The more posterior one is the **suprascapular nerve**, *which performs one of the four "tricks" (i.e., the "superior trick") mentioned above, passing through the suprascapular notch of the scapula to innervate the posterior pectoral girdle muscles* **infraspinatus** and **supraspinatus** (Plate 4.7b).

No significant nerves branch directly from the level of the six divisions of the brachial plexus, so the remaining seven nerves branch from the level of the three cords (Figure 4.3). The **medial cord** logically gives rise to the **medial pectoral nerve** (to both pectoralis major and minor), the **medial**

cutaneous nerve of the arm, and the **medial cutaneous nerve of the forearm**. The lateral cord logically gives rise to the **lateral pectoral nerve** (only to pectoralis major), and does not give rise to lateral cutaneous nerves of the arm and forearm because those of the forearm derive from the musculocutaneous nerve that bifurcates from this cord, while those of the arm derive from the axillary and radial nerves that bifurcate from the posterior cord (see Section 4.2.1.1). The **upper, middle, and lower subscapular nerves** branch from the posterior cord and *perform the fourth and final "trick" (i.e., the "inferior trick") listed above, running inferiorly and passing medially to the scapula*, to innervate the group formed by the **subscapularis**, **teres major**, and **latissimus dorsi**. At the level of the scapula, the upper and lower subscapular nerves innervate the subscapularis. Then the middle (thoracodorsal) and lower nerves continue their descent inferiorly until they pass the inferior portion of the scapula to reach the more inferior and posterior (dorsal) muscles latissimus dorsi and teres major, respectively, thence the name "inferior trick" (Plate 4.7b) (Table 4.1).

4.2.1.2 Nerves of the Arm, Forearm, and Hand

Because the axillary nerve only provides motor innervation for the teres minor and deltoid, it does not innervate muscles of the arm, forearm, and hand. Therefore, we are left with four nerves—**musculocutaneous nerve**, **median nerve**, **ulnar nerve**, and **radial nerve**—to provide motor innervation to all the muscles of these three anatomical regions (Figure 4.3). The radial nerve and its branches—the **deep radial nerve**, which continues as the **posterior interosseous nerve**—innervate all the posterior muscles of the arm and forearm, as might be expected based on their posterior location. They do not innervate muscles of the hand, because the hand has no

---

BOX 4.6 EASY WAYS TO REMEMBER: MOTOR INNERVATION OF ANTERIOR ARM MUSCLES

Remember that the *musculocutaneous nerve* does all the work at the arm level, innervating the three anterior arm muscles; the *median nerve* and *radial nerve* do most of the work in the anterior and posterior forearm, respectively; the *ulnar nerve* thus does most of the work in the hand, innervating most of the intrinsic hand muscles.

true posterior (dorsal) intrinsic muscles, only anterior (ventral) ones, as will be explained in Section 4.2.

Keeping in mind the overall organization just discussed, it becomes relatively easy to learn the details (Figure 4.3). After providing motor innervation to the three anterior arm muscles—**biceps brachii, coracobrachialis**, and **brachialis**—the musculocutaneous nerve gives rise to the **lateral cutaneous nerve of the forearm**. The **radial nerve** runs through the **radial (spiral) groove of the humerus** and provides motor innervation to the posterior (extensor) muscles of the arm. It then divides, at the level of the elbow, into the **deep branch of the radial nerve** and the **superficial branch of the radial nerve**. The superficial branch is purely sensory; it follows the brachioradialis muscle toward the wrist, and then becomes subcutaneous at the level of the wrist to innervate the skin on the dorsum of the thumb and lateral wrist. The deep branch of the radial nerve, and its continuation after crossing deep to the supinator muscle—the **posterior interosseous nerve**, so named because it runs posteriorly (dorsally) to the **interosseous membrane**—provide motor innervation for all the posterior muscles of the forearm. At the level of the elbow, the **median nerve** also divides into two branches: One is the continuation of the median nerve itself that travels between the deep and superficial muscles of the anterior forearm; the other is named the **anterior interosseous nerve** because it runs anteriorly (ventrally) along the interosseous membrane. Therefore, a major difference between the anterior and posterior interosseous nerves, which is often not emphasized enough to students, is that the former is a *branch* of the median nerve, while the latter is the *continuation* of the deep branch of the radial nerve.

The median nerve and the anterior interosseous nerve derived from it innervate all anterior (ventral) muscles of the forearm, with two expections: (1) the **flexor carpi ulnaris** and (2) the part of the **flexor digitorum profundus** that flex digits 4 and 5; these structures receive instead innervation from the ulnar nerve (see Section 4.3.3). This is logical because these structures lie in the medial (ulnar) region of the upper limb, therefore too far from the centrally located median nerve.

Regarding the innervation of hand muscles by the ulnar nerve, there are two exceptions: (1) the thenar muscles flexor pollicis brevis, abductor pollicis brevis, and opponens pollicis, and (2) the lumbricals 1 and 2. In the distal forearm, the median nerve gives rise to a **palmar cutaneous branch** that receives sensory innervation from the skin of the central palm. The median nerve enters the hand by passing inside the **carpal tunnel** (enclosed by the **flexor retinaculum**); see notes in Section 4.3.3 about **carpal tunnel syndrome**. After passing through this tunnel, the median nerve gives rise to the **recurrent branch of the median nerve** (innervating the muscles of the thenar compartment; i.e., base of the thumb: **flexor pollicis brevis**, **opponens pollicis**, and **abductor pollicis brevis**). Finally, the median nerve—which also supplies motor innervation to lumbricals 1 and 2—gives rise to the **common palmar digital nerves** and the **proper palmar digital nerves** that supply cutaneous innervation to the radial three and a half digits of the palmar side and the distal ends of digits 1, 2, 3, and half of 4 on the dorsum of the hand (Figure 4.3). The fact that the median nerve innervates the lumbricals 1 and 2 is logical: As said above, the median nerve innervates the part of the flexor digitorum longus going to digits 2 and 3—therefore, the lumbricals originating from the tendons of this muscle going to digits 2 and 3 (i.e., lumbricals 1 and 2) are logically also innervated by the medial nerve. Accordingly, it is logical that lumbricals 3 and 4 are innervated by the ulnar nerve, because these two lumbricals are originated by the tendons of the flexor digitorum profundus to digits 4 and 5, which are innervated by the ulnar nerve, as explained above.

At the distal forearm, the ulnar nerve passes inside the **Guyon's canal** and then gives off the **superficial branch of the ulnar nerve**, which supplies cutaneous innervation to digit 5 and the medial side of digit 4 as well as motor innervation to the muscle **palmaris brevis**. The **deep branch of the ulnar** nerve supplies motor innervation to all other hand muscles except those innervated by the median nerve.

BOX 4.7 EASY WAYS TO REMEMBER: INTEROSSEOUS VESSELS AND NERVES

Among the anterior and posterior interosseous neurovascular structures, the nerves are more logically named than the arteries. The radial and median nerves—which mainly run in the *center (midline) of the posterior and anterior regions, respectively,* of the distal arm/proximal forearm—give rise to the likewise *centrally located posterior and anterior interosseous nerves of the forearm,* respectively. On the contrary, both the **posterior interosseous artery** and **anterior interosseous artery** derive instead from the **ulnar artery** (via the **common interosseous artery**), which runs on the antero-medial region of the proximal forearm. Although spatially this configuration of the arteries seems illogical, it is necessary because no major arteries enter the proximal region of the forearm in the center of the anterior (ventral) and posterior (dorsal) side of the limb, as do the median and radial nerves. Here is a trick to remember that it is the ulnar artery that gives rise to the posterior and anterior interosseous arteries: There are three names used for the major neurovascular structures at the level of the forearm—radial, median, and ulnar, and each one is the name of a structure that gives rise to interosseous structures. The *radial* nerve (logically) gives rise to the posterior interosseous nerve; the *median* nerve (also logically) gives rise to the anterior interosseous nerve; so, the *ulnar* artery must give rise to both the anterior and posterior interosseous arteries.

4.2.1.3 Cutaneous Nerves of the Upper Limb

Students of gross anatomy do not often see most of the **cutaneous nerves of the upper limb** and thus educators tend not to focus so much on these nerves (Plate 4.4). However, a good knowledge of the cutaneous nerves of the limb is clinically important, and enriches the student's understanding of the deeper nerves, and therefore of the overall organization of the upper limb as a whole. For instance, students often memorize that the **medial cord of the brachial plexus** gives rise to the **medial cutaneous nerve of the forearm** and **medial cutaneous nerve of the arm** (Figure 4.3). They are also told that the lateral cord of the brachial plexus does not give rise directly to cutaneous nerves, because the **lateral cutaneous nerve of the forearm** is a continuation of the **musculocutaneous nerve** derived from this lateral cord. So, what about the lateral skin of the arm, does it not have cutaneous innervation? If it does, where does it come from?

There are actually two cutaneous nerves that supply the lateral skin of the arm: The **superior lateral cutaneous nerve of the arm** comes from the **axillary nerve**, while the **inferior lateral cutaneous nerve of the arm** comes from the **radial nerve**, so both branch from nerves derived from the posterior cord of the brachial plexus (Figure 4.3). The **posterior cutaneous nerve of the arm** and the **posterior cutaneous nerve of the forearm** both branch from the radial nerve, which makes sense because they mainly lie on the posterior side of the arm and forearm. The intercostobrachial nerve comes from cutaneous branches of **intercostal nerves**, which also come from ventral rami (see Section 6.2). Lastly, the medial, anterior, and dorsal portions of the wrist are innervated by the **superficial branch of the ulnar nerve**, median nerve, and radial nerve.

4.2.2 Upper Limb Blood Vessels

Angiogenesis is the process of blood vessel formation. In limb buds, this process occurs through a combination of individual angioblast (blood vessel precursor cell) movements throughout the lateral and paraxial mesoderm, followed by the assembly of these cells into endothelial cords and vesicles, which coalesce and hollow-out to form patent (open) channels. Initially, branches from several intersegmental arteries penetrate the limb buds, but most of these atrophy leaving a single, central artery bringing blood to early limbs. Smaller arterioles and capillaries radiate outwards from this central artery and are collected by a venous plexus located beneath surface ectoderm. Later, as muscles and cartilages begin to form, the adult pattern of venous return is established.

In general the arteries and veins of the human upper limb have similar configurations and analogous names. Therefore, gross anatomy students are usually encouraged to focus more on the arteries and to study in detail only the veins that differ markedly from the arteries. The **superficial veins of the upper limb**, and particularly of the forearm and arm, are examples of veins that have *no* clear analogs in the arterial system (Plate 4.4). The **axillary vein** is the major deep vein at the level of the axillary (or armpit) region. It is joined by the cephalic vein, so named for its direction of drainage toward the head, within the **deltopectoral triangle** (Plate 4.5b). Apart from these connections with the axillary vein, the superficial veins are also connected to the other deep veins via the **perforating veins**. The **cephalic vein** then continues superficially to pass into the **deltopectoral groove** between the deltoid and pectoralis major muscles. In the **cubital fossa** of the elbow region, the **median cubital vein** connects the cephalic vein and the **basilic vein** (which emerges from beneath the brachial fascia just proximal to the elbow and forms a network of veins surrounding the brachial artery). The boundaries of the cubital fossa are: lateral (brachioradialis), medial (pronator teres), superior (imaginary line connecting medial and lateral humeral epicondyles), superficial (antebrachial fascia), and deep (brachialis and supinator). The basilic and cephalic veins anastomose in the hand via the **dorsal venous arch**, which collects venous drainage from the posterior (dorsal) surface of the hand and digits.

The **axillary vein** is located within the **axillary sheath**, a connective tissue sleeve that also surrounds the **axillary artery** and the brachial plexus. The axillary artery (Figure 4.4) is the continuation of the **subclavian artery**; it begins at the lateral border of the first rib, and ends at the inferior border of the teres major where its name changes to **brachial artery** ("artery of the arm"). The easiest way to learn the branches of the axillary artery is to divide it into three parts: The first, second, and third parts lie medially, posteriorly, and laterally to the pectoralis minor muscle, respectively (Plate 4.9). Details about the specific muscles that the branches of the axillary artery supply arc given in Tables 4.1 and 4.2. The **first part of the axillary artery** has one branch: the **superior thoracic artery**, which makes sense because this is the most medial—and thus *superior*—branch of the axillary artery and lies in the *thoracic* region, supplying mainly the first and second intercostal spaces.

The **second part of the axillary artery** has two branches: the **thoracoacromial trunk** and the **lateral thoracic artery**. The name "thoracoacromial trunk" refers to its large area of distribution, from the *thoracic* region to the *acromion* region. This trunk has an **acromial branch** going mainly to the region of the acromion, a **deltoid branch** accompanying the cephalic vein in the deltopectoral groove, a **pectoral branch** supplying the pectoralis major and minor, and a **clavicular branch** supplying the subclavius. The lateral thoracic artery is named logically, as this artery is *lateral* to the pectoralis minor—the landmark for the second part of the axillary artery—and supplies blood to the *thoracic* region, namely, to the pectoralis major and minor and to the lateral thoracic wall.

The **third part of the axillary artery** has three branches: the **subscapular artery**, the **anterior circumflex humeral artery**, and the **posterior circumflex humeral artery**. As its name indicates, the subscapular artery runs inferiorly near the *subscapular* nerves. It then divides into the **thoracodorsal artery**—which, like the thoracodorsal (or middle subscapular) nerve, goes to the latissimus dorsi—and the **circumflex scapular artery** that goes to muscles of the posterior surface of the scapula. The anterior and posterior circumflex humeral arteries course, respectively, anteriorly and posteriorly to the surgical neck of the humerus, the latter artery passing through the quadrangular space of the back together with the axillary nerve (Plate 4.7b). Why does the body have circumflex blood vessels that surround a skeletal structure (e.g., the scapula or humerus) to meet with their counterparts? The answer is that such a **circulatory anastomosis** (connection)—between arteries or between veins provides a backup route for the flow of blood if one route is blocked or compromised.

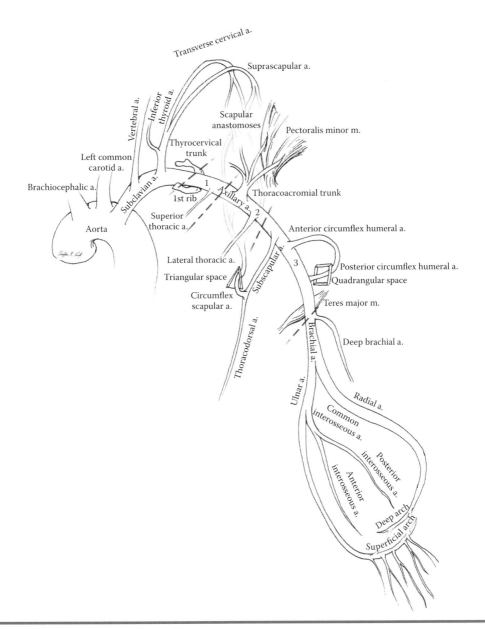

Figure 4.4 **Diagram of the arteries of the upper limb, schematic view. Note that only some branches of the subclavian artery are shown; other branches, such as the costocervical trunk and dorsal scapular and internal thoracic arteries are described, and shown in figures of Chapters 3 and 6 and the respective plates that go with them. (Modified from learning tool developed by the VCU School of Medicine, Class of 2009.)**

Having learned the branches of the axillary artery, it becomes much easier to understand the configuration of the arm and forearm arteries. The continuation of the axillary artery that enters the arm to provide supply to all the arm, forearm, and hand muscles is the **brachial artery**, which runs on the anterior side of the arm (Plate 4.3). If the body were free from evolutionary and developmental constraints, it would be more logical to have two major arteries, one for each of the two compartments of the arm (anterior and posterior). As this is not the case, the brachial artery has to perform a "trick" to supply to the triceps brachii and the other tissues of the posterior arm: It sends a proximal branch into the posterior arm compartment, as the **deep brachial artery** (or **deep artery of the arm**, or **profunda brachii artery**) (Figure 4.4). This artery courses around the posterior surface of the humerus, where it accompanies the radial nerve in the spiral (radial) groove, and then gives rise to the **radial collateral artery** that anastomoses with the **radial recurrent artery** branching from the radial artery (Plate 4.9b). The brachial artery then branches again, more distally, in the middle third of the arm, to give rise to the **superior ulnar collateral artery** and the **inferior ulnar collateral artery**. These two collateral arteries anastomose, respectively, with the **posterior ulnar recurrent artery** and the **anterior ulnar recurrent artery** branching from the ulnar artery.

In the region of the elbow, the brachial artery divides into the **radial artery** and the **ulnar artery** (Figure 4.4). Because the two latter arteries are mainly anterior arteries and the branches of the deep brachial artery are connected distally to recurrent arterial branches at the elbow region, another "trick" is required to provide blood supply to the posterior forearm compartment. Therefore, the ulnar artery gives rise to a branch—the **common interosseous artery**—that divides into the **anterior interosseous artery** and the **posterior interosseous artery** (Plate 4.9b). In Section 4.2.1.2, we explained in detail the configuration of the posterior and anterior interosseous neurovascular structures, and provided an easy way for students to remember that it is the ulnar artery that thus gives rise to the posterior and anterior interosseous arteries (Box 4.7).

The continuation of the ulnar artery then forms, distally, the **superficial palmar arch**, which gives rise to the **common digital arteries** that in turn give rise to the **proper digital arteries** (Figure 4.4). The continuation of the radial artery forms, distally, the **deep palmar arch**, which is connected via the **superficial palmar branch of the radial artery** to the superficial palmar arch (Plates 4.17b and 4.18). A clinically important feature of the radial artery is that it passes through the **anatomical**

snuffbox—enclosed by the tendons of the extensor pollicis brevis and extensor pollicis longus (Plate 4.15)—to reach the dorsum of the hand. Then the artery returns to the palmar surface by passing between the heads of the 1st dorsal interosseous muscle. In the palm, it then gives rise to the **deep palmar arch** and to a distinct artery for the thumb: the **princeps pollicis artery**.

4.3 Upper Limb Muscular System

The easiest way to understand the muscular structures of the adult human upper limb is to first learn their evolutionary and developmental origin. This method makes much more sense and dramatically reduces the amount of information that students need to learn/memorize by grouping the muscles into real, and thus logical, biological and anatomical units. While functional groups such as the "**rotator cuff**" are important in clinical contexts, such groups should be learned only *after* one learns the muscular system as a whole. The field of evolutionary developmental biology has been particularly helpful in illuminating the evolutionary history and ontogeny of the upper limb muscles, which are much better known today than when the first gross anatomy atlases and textbooks were written. Therefore, it is especially important to incorporate the results of recent publications when teaching or learning the anatomy of this region, to decrease the gap that exists between the current state of scientific knowledge and what is being taught in classrooms.

In response to signals from **limb bud** mesenchyme (Box 4.2), appendicular myoblasts bud off the lateral margin of nearby **somites** and move a short distance into the limb bud to segregate into separate *dorsal* and *ventral muscle condensations*. As the limb grows, subpopulations of myogenic cells within each of these condensations separate from one another to form individual muscles (Plate 4.2). The two masses give rise largely—but not exclusively—to the "true" extensor and flexor muscles of the limbs, respectively (as explained in Section 2.3, in an evolutionary/developmental sense, the true extensors of the upper limb correspond to the posterior arm and forearm muscles in human anatomical position, but due to the anatomical position of the lower limb in most textbooks the "true" extensors of the lower limb are actually shown in the anterior side of the thigh and leg). Coincident with the onset of individual muscle differentiation, the precursors of tendons become localized at sites corresponding to the future attachment sites of individual muscles; experiments have shown that these pretendon sites determine the orientation of muscle fibers.

Recent evolutionary and developmental studies have supported the idea that the **upper limb musculature** includes five main groups of muscles: the **axial muscles of the pectoral girdle (serratus anterior, rhomboid major, rhomboid minor, levator scapulae,** and **subclavius)** and the **appendicular muscles of the pectoral girdle (pectoralis major, pectoralis minor, infraspinatus, supraspinatus, deltoid, teres minor, subscapularis, teres major, and latissimus dorsi), arm, forearm,** and **hand** (Tables 4.1 through 4.7). The appendicular musculature of the pectoral girdle, arm, forearm, and hand essentially corresponds to the **abaxial musculature** of current evolutionary developmental biology. The axial pectoral girdle musculature is derived from the postcranial axial musculature, and is included in the **primaxial musculature**.

Developmental studies carried out in recent years have revealed important differences in the expression patterns associated with morphogenesis of the upper limb versus that of the lower limb and between proximal and distal regions within each limb. The formation of the muscles of the pectoral girdle occurs through mechanisms that are markedly different from the well-studied migration of myogenic cells from the somites to the limb bud muscles, which forms the arm, forearm, and hand. Instead, the abaxial pectoral girdle muscles develop by an "in-out" mechanism whereby migration of myogenic cells from the somites into the **limb bud** is followed by their extension from the proximal limb bud back onto the **thorax**. The primaxial pectoral girdle muscles are induced by the upper limb field that promotes myotomal extension (the migration of cells) directly from the somites. The appearance of the upper limb is followed by the development of the pectoral girdle that attaches the limb to the **axial skeleton**. The mechanisms involved in limb development are thus able to induce and recruit axial structures (e.g., part of the scapula and the primaxial pectoral girdle muscles) for the anchorage of the limb.

4.3.1 Muscles of Pectoral Girdle

To understand the configuration of the muscles of the pectoral girdle, you must recognize that the human body is not a perfect machine but the result of **evolutionary and developmental constraints**. As an example, we are "stuck" with four extremities, because we came from tetrapod ancestors. In a similar way, we are "stuck" innervating several back (dorsal) muscles via ventral rami (Plate 4.6). These muscles—deltoid, teres minor and major, latissimus dorsi, subscapularis, infraspinatus, supraspinatus, rhomboid major and minor, and levator scapulae—have shifted during evolution and development from the ventral to the dorsal side of the body (i.e., mainly from the pectoral region to the back), taking with them the distal portion of the nerves that innervate them. Because these muscles form four developmental and evolutionary groups (Table 4.1), there are accordingly four "*tricks*" that the nerves innervating each group have to perform to pass from the anterior (ventral) side of the body where the branchial plexus lies to the posterior (dorsal) side of the body where the muscles that they innervate are located. These four tricks are shown in Plate 4.7b and were listed and described in Section 4.2.

One advantage of knowing these tricks is that you will gain an appreciation of those aspects of human anatomy that do not seem to make sense—mostly related to evolutionary constraints—and are more prone to injury. For instance, because the axillary nerve has to pass into the **quadrangular space** (see Section 4.2) to go to the back of the body to innervate the deltoid and teres minor, this nerve is often affected in cases of surgical neck fracture or shoulder dislocation. The other three groups of pectoral muscles—one including the pectoralis major and minor, another the subclavius, and the other the serratus anterior—lie on the anterior (ventral) side of the body and therefore the nerves that supply them do not perform such "tricks."

4.3.1.1 *Posterior Muscles of Pectoral Girdle*

The **levator scapulae, rhomboid major**, and **rhomboid minor** are closely related developmentally and evolutionarily, being innervated by the **dorsal scapular nerve** (Plates 4.6 and 4.7; Table 4.1). The common function of these three muscles is to medially rotate the scapula. This rotation moves the inferior angle of the scapula medially, and that contributes 1° for each 2° that the arm is adducted. Therefore, although none of these three muscles directly attaches to the arm, they all contribute, indirectly, to the adduction of the arm. Apart from this function, the more horizontal orientation of the rhomboids allows them to also adduct (and therefore retract) the scapula, while the vertical orientation of the levator scapulae enables it to elevate the scapula (Table 4.1). The group formed by the **infraspinatus** and **supraspinatus** is innervated by the **suprascapular nerve** (Table 4.1).

BOX 4.8   EASY WAYS TO REMEMBER:
SUPRASCAPULAR NERVE AND ARTERY

A commonly used way to remember that the suprascapular nerve passes through the notch inferior to the **superior transverse scapular ligament** and that the suprascapular artery passes superior to ("above") this ligament is: "The Army passes over the bridge, while the Navy passes below the bridge."

In primates, including humans, the supraspinatus lies mainly on the superior side of the scapula and shoulder joint. The superior position of the supraspinatus within the shoulder girdle makes it well suited to abduct the arm when the muscle contracts (in particular of the first 15° of abduction, the remaining abduction being mainly performed by or in concert with the deltoid). Therefore, although the supraspinatus muscle is often grouped in the "**rotator cuff**" functional group with the subscapularis, infraspinatus, and teres minor, it is not a major shoulder rotator in humans. Unlike the supraspinatus, the infraspinatus in humans lies in a position similar to that seen in more generalized quadrupedal mammals, running from the posterior aspect of the scapula to the posterior proximal humerus and thus being a strong lateral rotator of the arm (Table 4.1; see Box 4.3).

Another developmental/evolutionary group includes the **subscapularis**, **teres major**, and **latissimus dorsi**, all innervated by **subscapular nerves** (Table 4.1). These muscles are actually often blended in human adults, the fibers of the subscapularis and teres major being blended proximally, and the tendons of the teres major and latissimus dorsi being blended distally.

BOX 4.9   EASY WAYS TO REMEMBER: FUNCTION OF
SUBSCAPULARIS, TERES MAJOR, AND LATISSIMUS DORSI

Think *"one step further"*: Each time you move further inferiorly in the body, you add one function. The common function of all three muscles is the medial (internal) rotation of the arm, because they all insert on the anterior surface of the humerus. Medial rotation is the sole function of the subscapularis. The teres major, inferior to the subscapularis, has the added function of adducting the arm. Lastly, the latissimus dorsi performs both these functions, plus extension of the arm (Table 4.1).

Table 4.1   Evolutionary/Developmental Groups of Posterior Muscles of the Pectoral Girdle

Group	Muscle	Nerve(s)	Function(s) in Common	Other Function(s)	Artery/Arteries	Origin	Insertion
Levator scapulae, rhomboid major and rhomboid minor (part of primaxial musculature)	Levator scapulae	Dorsal scapular n.	Medial rotation of scapula	Elevation of scapula	Dorsal scapular a.	Transverse processes of C1–C4	Medial border of scapula
	Rhomboid major			Adduction (and thus also some retraction) of scapula	Dorsal scapular a.	Spinous processes of T2–T5	
	Rhomboid minor				Dorsal scapular a./ deep branch of transverse cervical a.	Nuchal ligament and spinous processes of C7–T1	
Subscapularis, latissimus dorsi and teres major (part of abaxial musculature)	Subscapularis	Lower and upper subscapular n.	Medial rotation of arm	–	Subscapular a.	Subscapular fossa	Lesser tubercle of humerus
	Teres major	Lower subscapular n.		Adduction of arm	Subscapular a. and circumflex scapular a.	Inferior angle of scapula	Medial lip of intertubercular groove of humerus

(Continued)

Table 4.1 (Continued)   Evolutionary/Developmental Groups of Posterior Muscles of the Pectoral Girdle

Group	Muscle	Nerve(s)	Function(s) in Common	Other Function(s)	Artery/Arteries	Origin	Insertion
	Latissimus dorsi	Middle subscapular (thoracodorsal) n.		Adduction and extension of arm	Thoracodorsal a. (branch of subscapular a.)	Spinous processes of vertebrae T7–L5, thoracolumbar fascia, iliac crest, inferior ribs and, often, inferior angle of scapula	Intertubercular groove of humerus
Supraspinatus and infraspinatus (part of abaxial musculature)	Supraspinatus	Suprascapular n.	–	Abduction of arm	Suprascapular a.	Supraspinous fossa of scapula	Superior facet of greater tubercle of humerus
	Infraspinatus			Lateral rotation of arm	Suprascapular a. and circumflex scapular a.	Infraspinous fossa of scapula	Middle facet of greater tubercle of humerus
Deltoid and teres minor (part of abaxial musculature)	Deltoid	Axillary n.	–	Abduction of arm	Thoracoacromial a., and anterior and posterior humeral circumflex a.	Acromion and spine of scapula, lateral third of clavicle	Deltoid tuberosity of humerus
	Teres minor			Lateral rotation of arm	Circumflex scapular a. and posterior humeral circumflex a.	Lateral border of scapula	Inferior facet of greater tubercle of humerus

a. = artery; n. = nerve.

The **deltoid** and **teres minor** are both innervated by the **axillary nerve** after it emerges from the quadrangular space (Table 4.1; Plate 4.7b). The **superior border** of the quadrangular space is the inferior border of the teres minor, the **lateral border** is the **surgical neck of the humerus**, the **medial border** is the long head of the **triceps brachii**, and the **inferior border** is the superior border of the teres major. This quadrangular space should not be confused with the two triangles of the back, the triangle of auscultation, and the lumbar triangle. The **triangle of auscultation** (Plate 4.6) is bounded by the latissimus dorsi, trapezius, and rhomboid major, and is clinically important as a site where relatively thin musculature makes it easier to hear the sounds produced by the lungs with a stethoscope. The **lumbar triangle** is bounded by the latissimus dorsi, external oblique, and iliac crest and, in rare cases, is the site of a lumbar hernia. In many mammals, the **deltoid** has three heads, which can sometimes be distinguished in humans; the anterior, lateral, and posterior fibers are respectively involved in flexion, abduction, and extension of the arm. However, contraction of the full muscle in humans mainly contributes to a single function: the abduction of the arm, particularly after the **supraspinatus** begins arm abduction (i.e., mostly after first 15°). The function of the teres minor—lateral rotation of the arm—is similar to one of the functions of the more posterior (dorsal) head of the deltoid of many mammals, so in a way the teres minor of humans kept one of the original functions of the deltoid complex.

4.3.1.2 Anterior Muscles of Pectoral Girdle

The **anterior muscles of the pectoral girdle** (Table 4.2) are often designated the "**muscles of the pectoral region**" or the "**anterior thoracoappendicular group of muscles**." They include the serratus anterior, subclavius, and pectoralis major and minor.

The **serratus anterior** (Plates 4.5b and 4.6) is innervated by the **long thoracic nerve**, which as its name indicates must extend a long way inferiorly from the brachial plexus to innervate this muscle. As the serratus anterior runs posteriorly from the anterior surface of ribs 1–8 all the way to the medial border of the scapula, its contraction causes protraction of the scapula. The other function of the muscle is to laterally rotate the scapula, that is, to move its inferior angle laterally, thus contributing to abduction of the arm. Two muscles laterally rotate the scapula: the serratus anterior and the trapezius, while three muscles medially rotate the scapula—the levator scapulae and rhomboid major and minor.

Table 4.2 Evolutionary/Developmental Groups of Anterior Muscles of the Pectoral Girdle

Group	Muscle	Nerve(s)	Function(s)	Artery/Arteries	Origin	Insertion
Serratus anterior (part of primaxial musculature)	Serratus anterior	Long thoracic n.	Lateral rotation and protraction of scapula	Lateral thoracic a. and thoracodorsal a.	Ribs 1–8 or 1–9	Medial margin of scapula
Subclavius (part of primaxial musculature)	Subclavius	N. to subclavius	Depression of clavicle	Clavicular branch of thoracoacromial trunk	1st rib and its cartilage	Subclavian groove of clavicle
Pectoralis major and pectoralis minor (part of abaxial musculature)	Pectoralis major	Lateral pectoral n. and medial pectoral n.	Flexion, adduction, and medial rotation of arm	Pectoral branch of the thoracoacromial trunk	Clavicle (clavicular head), sternum and ribs (sternocostal head) and mainly aponeurosis of external oblique muscle (abdominal head)	Lateral lip of intertubercular groove of humerus
	Pectoralis minor	Medial pectoral n.	Depression and protraction of scapula		Ribs 3–5	Coracoid process of scapula

a. = artery; n. = nerve.

The **subclavius** is innervated by the **nerve to subclavius**, and its function is to depress the clavicle.

The **pectoralis major** and **pectoralis minor** have similar evolutionary and developmental origins, being innervated by the **lateral pectoral nerve** and the **medial pectoral nerve**. Pectoralis major, being larger, is innervated by both nerves, while the smaller pectoralis minor is innervated only by the medial pectoral nerve. The pectoralis major has a **clavicular head**, a **sternocostal head**, and an **abdominal head**. In anatomical position, the pectoralis major lies mainly anterior and medial to its insertion point onto the humerus, so its contraction adducts and flexes the arm. As its insertion is onto the lateral lip of the intertubercular groove, on the anterior surface of the humerus, it also medially (internally) rotates the arm. Because the pectoralis minor originates from ribs 3–5 and runs superiorly and posteriorly to insert onto the coracoid process of the scapula, its contraction causes the protraction and depression of the scapula. The deep fascia on the surface of the pectoralis major is called **pectoral fascia**, and it is continuous with the **axillary fascia**. Between the pectoralis major and the deltoid lies the **deltopectoral triangle** and the **cephalic vein**, which pierces the **clavipectoral fascia** and passes through the **costocoracoid** membrane (Plate 4.5b).

4.3.2 Muscles of Arm

The **brachial fascia** or **deep fascia of the arm** is formed by connective tissue that is continuous proximally with the **pectoral fascia** and the **axillary fascia** and distally with the **antebrachial fascia** or **deep fascia of the forearm**. Importantly, it is also connected to the medial and lateral sides of the humerus by intermuscular septa, forming the **posterior (extensor) compartment of the arm** and the **anterior (flexor) compartment of the arm**. The anterior compartment includes the muscles **biceps brachii**, **brachialis**, and **coracobrachialis** (Plate 4.10), which are innervated by the **musculocutaneous nerve** and form a developmental and evolutionary unit (Table 4.3). Students often confuse the **long and short heads of the biceps**; however, it is easy to remember once you realize that the long head is actually longer because it originates from the supraglenoid tubercle of the scapula, while the short head originates from the inferior tip of the coracoid process of the scapula. The posterior compartment includes the **triceps brachii** (Plate 4.11), which has a long head, a lateral head, and a medial head, and is innervated mainly by the **radial nerve** (Table 4.4). Most atlases and textbooks state that the anconeus is a posterior arm muscle

Table 4.3 Evolutionary/Developmental Groups of Anterior Muscles of the Arm

Group	Muscle	Nerve(s)	Function(s)	Artery/Arteries	Origin	Insertion
Biceps brachii, coracobrachialis and brachialis	Biceps brachii	Musculocutaneous n.	Flexion and supination of forearm	Brachial a.	Supraglenoid tubercle (long head) and coracoid process (short head) of scapula	Radial tuberosity (tendon) and deep fascia of medial forearm (bicipital aponeurosis)
	Coracobrachialis		Flexion and adduction of arm		Coracoid process of scapula	Medial side of humerus
	Brachialis		Flexion of forearm	Radial recurrent a.	Anterior surface of humerus	Coronoid process and tuberosity of ulna

Note: All are abaxial muscles.

a. = artery; n. = nerve.

Table 4.4 Evolutionary/Developmental Groups of Posterior Muscles of the Arm

Group	Muscle	Nerve(s)	Function(s)	Artery/Arteries	Origin	Insertion
Triceps brachii	Triceps brachii	Radial n. mainly, but also axillary n. for long head	Extension of forearm	Deep brachial a.	Infraglenoid tubercle of scapula (long head) and proximal (lateral head) and distal (medial head) to radial sulcus of humerus	Olecranon process of ulna

Note: All are abaxial muscles.

a. = artery; n. = nerve.

developmentally related to the triceps brachii, but it is actually a posterior *forearm* muscle that is developmentally and evolutionary closely related to the **extensor carpi ulnaris** (see Section 4.3.3).

4.3.3 Muscles of Forearm

The **anterior (flexor) compartment of the forearm** includes a **superficial group of forearm flexor muscles** and a **deep group of forearm flexor muscles** (Plate 4.13). The functions, attachments, innervation, and arterial blood supply of these muscles is given in Table 4.5. Therefore, in this paragraph, we will mainly refer to ways for the students to better understand, group, and study these muscles. For instance, textbooks and atlases of human gross anatomy often refer to an intermediate layer of forearm muscles, but this layer has no true developmental and evolutionary—or even anatomical—support. An easier way to study both the superficial and deep layers of forearm anterior (ventral) muscles is to go from lateral (radial) to medial (ulnar). Also, remember that all forearm flexors are innervated by the **median nerve** or its branch, the **anterior interosseous nerve**, with only two exceptions: The flexor carpi ulnaris and the part of the flexor digitorum longus that move digits 4 and 5 are innervated by the ulnar nerve. Even these exceptions make sense, because they comprise the most medial (ulnar) muscle structure of the superficial layer—the **flexor carpi ulnaris**—and the most medial part of a muscle of the deep layer—the ulnar half of the **flexor digitorum longus**, going to digits 4 and 5. They thus lie much closer to the ulnar nerve than to the more centrally placed median nerve, as explained above.

Within the **superficial group of anterior (ventral) forearm muscles** (Plate 4.13a), the most lateral muscle is the **pronator teres**, which can be easily recognized because it is the only superficial muscle that is not parallel to the main proximo-distal axis of the forearm and that does not reach the hand. Instead, this muscle runs obliquely—that is, disto-laterally—from the **medial epicondyle of the humerus**—the proximal attachment of the **common flexor tendon of most anterior forearm muscles**—to attach onto the shaft of the radius. The pronator teres is thus, logically, a flexor and *pronator* of the forearm. Medially to the pronator teres lies the **flexor carpi radialis**: As its name indicates, it goes to the *radial* (lateral) side of the carpal region, and thus flexes and abducts the hand. Medially to the flexor carpi radialis lies the **palmaris longus**: As its name indicates, this is a *long*, thin muscle that inserts onto the **palmar fascia (palmar**

Table 4.5 Evolutionary/Developmental Groups of Anterior (Ventral) Muscles of the Forearm

Group	Muscle	Nerve(s)	Function(s)	Artery/Arteries	Origin	Insertion
Superficial anterior (ventral) forearm muscles	Pronator teres	Median n.	Flexion and pronation of forearm	Ulnar a. and radial a.	Medial epicondyle of humerus (humeral head) and coronoid process of ulna (ulnar head)	Body of radius
	Flexor carpi radialis		Flexion and abduction of hand	Ulnar a.	Medial epicondyle of humerus	Bases of metacarpals 2 and 3
	Palmaris longus		Weak flexion of hand			Palmar aponeurosis
	Flexor digitorum superficialis		Flexion of hand and digits 2, 3, 4, and 5 (middle phalanges, so at proximal interphalangeal joints)		Medial epicondyle of humerus and parts of radius and ulna	Middle phalanges of digits 2, 3, 4, and 5
	Flexor carpi ulnaris	Ulnar n.	Flexion and adduction of hand		Medial epicondyle of humerus and olecranon of ulna	Pisiform, hook of hamate, base of metacarpal 5

(Continued)

Table 4.5 (Continued) Evolutionary/Developmental Groups of Anterior (Ventral) Muscles of the Forearm

Group	Muscle	Nerve(s)	Function(s)	Artery/Arteries	Origin	Insertion
Deep anterior (ventral) forearm muscles	Pronator quadratus	Median n. (anterior interosseous n.)	Pronation of forearm	Anterior interosseous a.	Distal ulna	Distal radius
	Flexor pollicis longus		Flexion of distal phalanx of thumb		Radius and interosseous membrane	Distal phalanx of thumb
	Flexor digitorum profundus	Median n. (anterior interosseous n.; part of muscle to digits 2 and 3) and ulnar n. (part of muscle to digits 4 and 5)	Flexion of hand and of digits 2, 3, 4, and 5 (distal phalanges, so at distal interphalangeal joints)		Ulna and interosseous membrane	Distal phalanges of digits 2, 3, 4, and 5

Note: All are abaxial muscles.

a. = artery; n. = nerve.

aponeurosis) and not directly onto bone—it is therefore a weak flexor of the wrist (see Box 4.10). Most of the fibers of the **flexor digitorum superficialis** lie medially, and deep, to the palmaris longus, but the two muscles are actually often deeply blended proximally. As its name indicates, the **flexor digitorum superficialis** flexes several digits—thence *"digitorum,"* and not "digiti." Because the **high mobility of the thumb** played a crucial role in the evolution of the human lineage, the long anterior and posterior muscles of the forearm that attach directly to various digits do not attach to the thumb: They attach only to digits 2, 3, 4, and 5. Humans have various peculiar muscles that are not found in other animals to exclusively flex, or extend, the thumb (see also notes about the full opposition of the thumb, in Section 4.1.3). Therefore, the flexor digitorum superficialis flexes digits 2, 3, 4, and 5: It extends to their middle phalanges, through tendons that bifurcate distally to form a bridge so the tendons of the flexor digitorum profundus can pass underneath to reach the distal phalanges of these four digits. The last—most medial—superficial anterior muscle of the forearm is the **flexor carpi ulnaris**, which, as explained above, is innervated by the ulnar nerve, being one of only two anterior forearm muscles *not* innervated by the median nerve or its branches. As its name indicates, the flexor carpi ulnaris goes to the *ulnar* (medial) side of the carpal region, and thus *flexes* and adducts the hand.

BOX 4.10   EVOLUTION, BIRTH DEFECTS, ANATOMICAL VARIATIONS, AND VESTIGIAL STRUCTURES: THE PALMARIS LONGUS

The case of the **palmaris longus** demonstrates how knowledge of the **evolutionary history of the human lineage** can help us to understand not only the gross anatomy but also **anatomical variations** of modern humans. In most primate species, the palmaris longus muscle is found in almost 100% of individuals. This makes sense because most primates rely heavily on powerful flexion of the hand; for instance, to grasp and navigate through the branches of the trees and avoid terrestrial predators. However, the palmaris longus is said to be present in only 80%–85% of the members of our species, *Homo sapiens*. Further, the muscle's frequency is highly variable within different geographic groups of this species, being higher than 85% in some groups and lower than 80% in others, meaning that the frequency of this muscle has changed, and continues to change, within the evolution of our lineage. Its overall shape is also changing: In many of those modern humans who do possess this muscle, it is a thin, weak structure—a **vestigial structure** in evolutionary terminology. In simplified terms, there is probably no **positive natural selective pressure** to lose the palmaris longus in our species, as in general it is not related to injuries or **pathologies** that increase the risk of early death or decrease the ability to have children. However, because having this muscle does not increase our chances of survival and reproduction, there is likely also no positive natural selection pressure to keep it either. Thus, the palmaris longus is functionally and evolutionarily "neutral," and as such it is free to vary widely in its presence/absence, shape, and size. Such evolutionarily "neutral" structures tend to be lost over long geological periods, because although there is no direct negative selective pressure against them, the energy used to form them could instead be used to form other structures that are more beneficial for the organism's survival. A classic example is the eyes of cavefish, which have been lost independently in many different cavefish lineages, simply because they are not advantageous (they are "neutral") to the fish's survival and fecundity. The palmaris longus seems to be following the same pattern in human evolution. Interestingly, evolutionarily "neutral" structures are often absent in individuals with severe **birth defects**. These defects should not be confused with **anatomical variations**, which refer exclusively to differences in form within the "normal" population—that is in individuals such as the ones you will be dissecting in the gross anatomy lab, who, apart from these variations, do not seem to have any birth defects. The absence of the palmaris longus in about 20%–15% of the normal human population, therefore, represents an emblematic example of anatomical variation.

There are only three muscles within the **deep group of anterior (ventral) forearm muscles**: flexor pollicis longus, flexor digitorum profundus, and pronator quadratus (Plate 4.13b). The **flexor pollicis longus** and the **flexor digitorum profundus** lie in the same antero-posterior level of the forearm, and both insert on the distal phalanges of the digits. The difference between these muscles is that, as the names indicate, the flexor pollicis longus inserts on the thumb ("pollex"), while the flexor digitorum profundus inserts on digits 2, 3, 4, and 5. The evolutionary history of these two muscles provides yet another example illustrating the crucial role played by the **high mobility of the thumb** in human evolution. In the vast majority of mammals, a single long flexor muscle inserts on the distal phalanges of all five digits. That ancestral muscle separated into two in humans, one inserting on digits 2–4 (flexor digitorum profundus) and the other inserting on digit 1, allowing this digit to be flexed independently from the other digits. As explained at the beginning of this section, the lateral part of the flexor digitorum profundus—that is, the part that sends tendons to digits 4 and 5—is one of the two anterior forearm muscle structures that are innervated by the ulnar nerve. Deep (posterior or dorsal) to the flexor pollicis longus and the flexor digitorum profundus lies the **pronator quadratus**, which as its name indicates is a *quadrangular* muscle that runs from distal ulna to the distal radius to *pronate* the forearm.

As explained in Section 4.1.3, the space between the **carpal bones** and the **flexor retinaculum** is designated the **carpal tunnel**. Many of the tendons of the anterior (ventral) forearm muscles pass through this tunnel, together with the **median nerve**: all the tendons of the **flexor digitorum profundus**, of the **flexor digitorum superficialis**, and of the **flexor pollicis longus**. The **common flexor sheath (ulnar bursa)** and three **digital synovial sheaths**, and the **synovial sheath of the flexor pollicis longus (radial bursa)** serve to decrease friction and increase mobility of the tendons of the long anterior (ventral) forearm muscles in the carpal tunnel and more distal regions of the hand. However, swelling of the common flexor sheath may cause compression of the median nerve in the carpal tunnel, resulting in **pain** and **paresthesia** ("pins and needles") of digits 1, 2, and 3 and weakness of the muscles of the thenar eminence (the muscular bulge at the base of the thumb; see Section 4.3.4). This condition is known as **carpal tunnel syndrome**.

The **posterior (extensor) compartment of the forearm** (Plates 4.14 and 4.15) includes muscles exclusively innervated by the radial nerve, the deep radial nerve that arises from it, or the distal continuation of the deep radial nerve: the posterior interosseous nerve. This compartment includes superficial and deep layers, and most of its muscles originate from the lateral epicondyle of the humerus (Table 4.6).

As explained in Section 4.3.2, although the **anconeus** assists the **triceps brachii** in forearm extension, developmentally and evolutionary, it is a posterior forearm muscle (Table 4.6) closely related to the extensor carpi ulnaris (Plate 4.14). The **anconeus** is therefore the most proximal of the **superficial posterior (dorsal) muscles of the forearm**. The **anconeus** inserts onto the olecranon process of the ulna and the annular ligament. It functions to

Table 4.6 **Evolutionary/Developmental Groups of Posterior (Dorsal) Muscles of the Forearm**

Group	Muscle	Nerve(s)	Function(s)	Artery/Arteries	Origin	Insertion
Superficial posterior (dorsal) forearm muscles	Anconeus	Radial n. (including deep radial n. for extensor carpi radialis brevis)	Extension of forearm	Deep brachial a. and recurrent interosseous a.	Lateral epicondyle of humerus	Olecranon process and adjacent portion of ulna; annular ligament
	Brachioradialis		Flexion of forearm	Radial recurrent a.	Lateral supracondylar ridge of humerus	Distal radius, including styloid process
	Extensor carpi radialis longus		Extension and abduction of hand	Radial a.	Lateral supracondylar ridge of humerus	Metacarpal 2
	Extensor carpi radialis brevis		Extension and abduction of hand		Lateral epiconcyle of humerus	Metacarpal 3
	Extensor digitorum		Extension of hand and of digits 2, 3, 4, and 5 (both the middle and distal phalanges)	Posterior interosseous a.	Lateral epiconcyle of humerus	Extensor expansion of digits 2, 3, 4, and 5 (therefore reaching both the middle and distal phalanges of each digit)
	Extensor digiti minimi		Extension of digit 5 (both the middle and distal phalanges)			Extensor expansion of digit 5 (therefore reaching both the middle and distal phalanges of this digit)

(Continued)

Table 4.6 (*Continued*) Evolutionary/Developmental Groups of Posterior (Dorsal) Muscles of the Forearm

Group	Muscle	Nerve(s)	Function(s)	Artery/Arteries	Origin	Insertion
	Extensor carpi ulnaris		Extension and adduction of hand	Ulnar a.	Lateral epicondyle of humerus and ulna	Metacarpal 5
Deep posterior (dorsal) forearm muscles	Supinator	Radial n. (deep radial n. for supinator and posterior interosseous n. for abductor pollicis longus, extensor pollicis, extensor pollicis longus and extensor indicis)	Supination of forearm	Radial recurrent a.	Lateral epicondyle, proximal ulna, and radial collateral and annular ligaments	Shaft of radius
	Abductor pollicis longus		Abduction and extension of metacarpal 1 and thus of thumb	Posterior interosseous a.	Ulna, radius, and interosseous membrane	Metacarpal 1
	Extensor pollicis brevis		Extension of proximal phalanx of thumb		Radius and interosseous membrane	Proximal phalanx of thumb
	Extensor pollicis longus		Extension of distal phalanx of thumb		Ulna and interosseous membrane	Distal phalanx of thumb
	Extensor indicis		Extension of digit 2 (both the middle and distal phalanges)			Extensor expansion of digit 2 (therefore reaching both the middle and distal phalanges of this digit)

Note: All are abaxial muscles.

a. = artery; n. = nerve.

constrain the axis of rotation of the forearm during supination so that it is aligned with the long axis of the radius and ulna (as in using a screwdriver). The anconeus originates from the **lateral epicondyle of the humerus**, which is also the origin of the **common extensor tendon of many posterior forearm muscles**. The long superficial posterior (dorsal) muscles, going from lateral to medial, are the **brachioradialis**, **extensor carpi radialis longus**, **extensor carpi radialis brevis**, **extensor digitorum**, **extensor digiti minimi**, and **extensor carpi ulnaris**. The **brachioradialis**, as its name indicates, runs from the arm (*brachium*) to the *radius*. This is a peculiar muscle because, despite being part of the posterior—and thus extensor—compartment of the forearm, it actually flexes the forearm. This action is possible because the fibers of the brachioradialis pass anterior (ventral) to the elbow joint, due to its origin above the supracondylar ridge and its insertion onto the anterior (ventral) surface of the distal radius. Therefore, this muscle is visible in an anterior (ventral) view of the forearm but should not be confused with the anterior forearm muscles. Medially to the **brachioradialis**, and fully on the posterior (dorsal) side of the forearm, lie the **extensor carpi radialis longus** and the shorter **extensor carpi radialis brevis**. As indicated by their names, they attach onto the *radial* side of the *carpal* region and thus *extend* and *abduct* the hand. The former muscle is *longer* than the latter because it originates from the lateral supracondylar ridge of the humerus, and not from the lateral epicondyle of this bone ("*brevis*" means short; think "brief," or "brevity is the soul of wit").

As its name indicates, the **extensor digitorum** inserts on various digits: As explained above, because of the high mobility and freedom of the thumb, the extensores and flexores digitorum in humans insert on digits 2, 3, 4, and 5, but not 1. Also, the simple name "**extensor digitorum**" indicates that, unlike the anterior compartment of the forearm that contains both a **flexor digitorum superficialis** and a **flexor digitorum profundus**, the posterior compartment contains only one long extensor muscle for digits 2, 3, 4, and 5. Logically, this single muscle is the antagonist of the two flexor digitorum muscles of the anterior compartment: That is, it performs the opposite action (extension vs. flexion) and can extend both the middle and the distal phalanges of digits 2, 3, 4, and 5. The extensor digitorum and all the other extensors of the forearm that attach onto **fingers** (digits 2 to 5) are able to reach the middle and distal phalanges because they are associated with the **extensor expansions** (or **dorsal expansions**, or **dorsal hoods**) attached to the phalanges of these digits (Plate 4.14). The **extensor digitorum** passes deep (anterior or ventral) to the **extensor retinaculum**, and its four distal tendons are tied together by **intertendinous connections** at the hand region. Just medially to the **extensor digitorum** lies the **extensor digiti minimi**, which as its name indicates inserts only onto digit 5 (the littlest finger, thence the designation "*minimi*"). Medially to this muscle lies the **extensor carpi ulnaris**, which as indicated by its name goes to the *ulnar* side of the *carpal* region and thus *extends* and *adducts* the hand.

BOX 4.11   EASY WAYS TO REMEMBER: DEEP POSTERIOR (DORSAL) MUSCLES OF THE FOREARM

The **deep posterior (dorsal) muscles of the forearm**, like the other compartments, include a more proximal, oblique muscle and various longer, more distal muscles. The proximal muscle of the deep forearm compartment is the **supinator**, which connects the distal humerus and the proximal ulna and surrounding (radial collateral and annular) ligaments and logically *supinates* the forearm. This muscle is an important landmark for neurovascular structures because it is after passing the distal border of this muscle that the **deep radial nerve** changes its name to **posterior interosseous nerve** (Figure 4.4).

To remember the functions of the other four deep posterior muscles of the forearm, from more lateral to medial, is to think: *one step further* (similar to the rule for the posterior pectoral girdle muscles in Box 4.9, but instead of adding a function you will keep the same function and add a joint). The first—most lateral—of these muscles is the **abductor pollicis longus**, which as its name indicates is a *long abductor* of the *thumb* (meaning that there is also a short thumb abductor, as will be seen in Section 4.3.4). The abductor pollicis longus is able to perform this function because it inserts on the antero (ventro)-lateral surface of metacarpal 1, thus abducting—and to a lesser extent also extending—the thumb, mainly at the **carpometacarpal joint** between the carpals and metacarpal 1. As explained in Section 4.1, this carpometacarpal joint, and to a lesser extent the one between the carpals and metacarpal 5, are the joints that allow **opposition**, or direct contact, of the palmar surfaces of digits 1 and 5 (Figure 4.2). Medially to the abductor pollicis longus lies the **extensor pollicis brevis**, which as its name indicates *extends the thumb* but inserts on the proximal phalanx of digit 1, going *one step further* than the abductor pollicis longus and thus extending the thumb at the **metacarpophalangeal joint**. Following this logic, it should be clear that the next—more medial—muscle, the **extensor pollicis longus**, will go *one step further*, to the distal phalanx of the thumb, to extend this phalanx at the **thumb interphalangeal joint**. Lastly, applying the same reasoning and knowing that there is only one interphalangeal joint in the thumb because it has only two phalanges, you can predict that the **extensor indicis** will *go one step further*, that is, to digit 2 instead. Because it inserts via the extensor expansion, this muscle attaches onto—and thus can extend—both the middle and distal phalanges of digit 2.

Digits 1, 2, and 5 each have two distinct extensor muscles: **extensor pollicis brevis and extensor pollicis longus** for digit 1; **extensor digitorum and extensor indicis** for digit 2; and **extensor digitorum and extensor digiti minimi** for digit 5. Most other mammals have two extensor muscles that attach to each of the fingers (digits 2 to 5) but only one muscle that extends the thumb. The highly mobile human thumb has an additional extensor muscle, while the fingers have two fewer extensors (namely, the deep extensors to digits 3 and 4 are normally missing in humans). It is the tendons of the extensor pollicis brevis and extensor pollicis longus that delimit, at the carpal region, the **anatomical snuffbox** (Plate 4.15). This depression was often used for placing and subsequently sniffing powdered tobacco, hence its name. For gross anatomy students, this surface is an important clinical landmark because deep to it lies the **radial artery** as well as the articulation between the radius and scaphoid, the latter of which is the most often fractured bone of the wrist.

4.3.4 Muscles of Hand

The **intrinsic hand muscles** are located only on the anterior (ventral) side of the limb; accordingly, they are exclusively innervated by the anterior (ventral) **median nerve** and **ulnar nerve** (Plates 4.17 and 4.18). Table 4.7 summarizes the innervation, functions, arterial blood supply, and attachments of these muscles.

The most superficial muscle of the hand is the **palmaris brevis** (Plate 4.17a), which is accordingly innervated by the **superficial branch of the ulnar nerve**. No other hand muscle is innervated by this branch. The palmaris brevis is the single muscle remaining from a group of very superficial flexor muscles present in our early tetrapod ancestors. It is essentially

Table 4.7 Evolutionary/Developmental Groups of Intrinsic (All Anterior or Ventral) Muscles of the Hand

Group	Muscle(s)	Nerve(s)	Function(s)	Artery/Arteries	Origin	Insertion
Lumbricals	Lumbricals 1, 2, 3, and 4	Median n. (lumbricals 1 and 2) and ulnar n. (lumbricals 3 and 4)	Flexion of proximal and extension of middle and distal phalanges of digits 2, 3, 4, and 5	Superficial and deep palmar arches, and dorsal and common palmar digital arteries	Tendons of flexor digitorum profundus	Base of proximal phalanges and extensor expansions of digits 2, 3, 4, and 5 (therefore reaching both the middle and distal phalanges of each digit)
Thenar muscles	Abductor pollicis brevis	Recurrent branch of median n. (but the	Abduction of proximal phalanx of thumb	Superficial palmar arch	Trapezium, scaphoid and transverse carpal ligament	Base of proximal phalanx of thumb
	Flexor pollicis brevis (superficial 1 and deep heads)	deep, and to a lesser extent the superficial, heads of the flexor pollicis	Flexion of proximal phalanx of thumb		Trapezium, flexor retinaculum	Proximal phalanx of thumb
	Opponens pollicis	brevis are often also innervated by deep branch of the ulnar n.)	Opposition of the thumb, at its carpometacarpal joint		Trapezium and transverse carpal ligament	Metacarpal 1

(Continued)

Table 4.7 (*Continued*) Evolutionary/Developmental Groups of Intrinsic (All Anterior or Ventral) Muscles of the Hand

Group	Muscle(s)	Nerve(s)	Function(s)	Artery/Arteries	Origin	Insertion
Palmaris brevis	Palmaris brevis	Superficial branch of the ulnar n.	Pulling of skin over hypothenar eminence	Palmar arches	Flexor retinaculum and palmar aponeurosis	Skin over hypothenar eminence
Hypothenar muscles	Abductor digiti minimi	Deep branch of ulnar n.	Abduction proximal phalanx of digit 5	Ulnar a.	Pisiform and surrounding structures	Base of proximal phalanx of digit 5
	Flexor digiti minimi brevis		Flexion of proximal phalanx of digit 5		Hamate	Base of proximal phalanx of digit 5
	Opponens digiti minimi		Opposition of digit 5, at its carpometacarpal joint		Hook of hamate and flexor retinaculum	Metacarpal 5
Adductors	Adductor pollicis (transverse and oblique heads)	Deep branch of ulnar n.	Adduction of proximal phalanx of thumb	Deep palmar arch	Metacarpal 3 (transverse head) and metacarpals 2 and 3 and adjacent carpal bones (oblique head)	Proximal phalanx of thumb

(*Continued*)

Table 4.7 (*Continued*) Evolutionary/Developmental Groups of Intrinsic (All Anterior or Ventral) Muscles of the Hand

Group	Muscle(s)	Nerve(s)	Function(s)	Artery/Arteries	Origin	Insertion
	Adductor pollicis accessorius (pollical palmar interosseous muscle of Henle)		Mainly unknown, might be insignificant or, for instance, weak adduction and/or flexion of proximal phalanx of thumb	Radial a. (perhaps from princeps pollicis a. that branches from it)	Base of metacarpal 1	Base of proximal phalanx of thumb and surrounding structures
Interossei	Dorsal interossei 1, 2, 3, and 4	Deep branch of ulnar n.	Abduct digits 2 and 3 to radial side and 3 and 4 to ulnar side	Deep palmar a.	Metacarpals	Radial sides of proximal phalanges of digits 2 and 3, and ulnar side of proximal phalanges of digits 3 and 4
	Palmar (volar) interossei 1, 2, and 3		Adduct digit 2 to ulnar side and digits 4 and 5 to radial side			Ulnar side of proximal phalanx of digit 2 and radial sides of proximal phalanges of digits 4 and 5

Note: All are abaxial muscles.

a. = artery; n. = nerve.

vestigial in humans (Box 4.10), and is often inadvertently destroyed during dissection as it lies just deep to the skin of the palm of the hand. Deep to this muscle, and to the tendons of flexor digitorum superficialis, at the center of the carpal and metacarpal region, lie the **lumbricals** (Plate 4.17b). These muscles have a peculiar dual function: They flex the proximal phalanges and extend the distal phalanges of the fingers. These two actions are possible because the **lumbricales** have two insertions: They insert onto the radial side of the anterior (ventral) region of the base of the proximal phalanges of digits 2, 3, 4, and 5 to flex the proximal phalanges, and they also extend distally to insert onto the extensor expansion of these digits, thus reaching—and extending—their middle and distal phalanges. The innervation of the lumbricals is far easier to learn than their function; they share an innervation with their proximal attachments, the tendons of the **flexor digitorum profundus**. Therefore, the lumbricals to digits 2 and 3 are innervated by the median nerve, and the lumbricals to digits 4 and 5 are innervated by the ulnar nerve (see innervation of flexor digitorum profundus in Section 4.3.3).

The muscles of the **thenar eminence** (Plate 4.17a) are all innervated by the **recurrent branch of the median nerve**. The **deep head of the flexor pollicis brevis** and to a lesser extent the **superficial head of the flexor pollicis brevis** are often also innervated by the **deep branch of the ulnar nerve**. All the other muscles of the hand are innervated by the deep branch of the ulnar nerve. The thenar eminence includes the **abductor pollicis brevis**, the **flexor pollicis brevis**, and the **opponens pollicis**, which logically are mainly involved, respectively, in the *abduction*, *flexion*, and *opposition* of the thumb. There is much confusion—mostly in the more specialized publications, but also in many textbooks and atlases of human gross anatomy—between the **deep head of the flexor pollicis brevis** and a very thin muscle that is often designated **adductor pollicis accessorius** (or, alternatively, **pollical palmar interosseous muscle of Henle**) (Figure 4.5). Recent detailed studies have shown that these are in fact different structures that are both *usually* present in humans (in 90% or more of cases). Therefore, it is now possible and advisable to include both muscles in all atlases and textbooks of human gross anatomy, as is done in Table 4.7. The adductor pollicis accessorius seems to be a derived feature of humans and a few primates (i.e., this trait is not present in other groups of animals), which might have had more significance at earlier stages of human evolution. This muscle now seems to be an evolutionarily "neutral" structure in humans because, although it is almost always present, it displays a huge

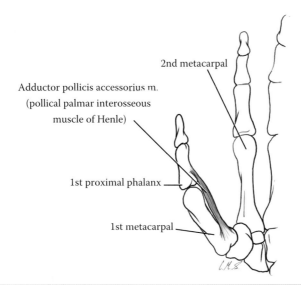

2nd metacarpal

Adductor pollicis accessorius m.
(pollical palmar interosseous
muscle of Henle)

1st proximal phalanx

1st metacarpal

Figure 4.5 **Adductor pollicis accessorius muscle in the left hand, palmar view.**

diversity of attachments and subdivisions. As its name indicates, the adductor pollicis accessorius is likely the result of differentiation of the **adductor pollicis** muscle, namely, its **oblique head**, so it is not part of the thenar eminence, as is the deep head of the flexor pollicis brevis.

The three muscles of the hypothenar eminence—**abductor digiti minimi, flexor digiti minimi brevis**, and **opponens digiti minimi** (Plate 4.18)—are basically mirror image of the three muscles of the thenar eminence (however, unlike the flexor pollicis brevis, the **flexor digiti minimi brevis** usually has only a single, undifferentiated head). Their functions, logically, are *abduction, flexion*, and *opposition* of digit 5, respectively. The movements of all joints of the digits, including adduction and abduction relative to the axis of the hand and the opposition of the thumb, were described in Section 4.1.3. Knowledge of these movements will help explain the configuration of the two deepest groups of hand muscles: the **palmar interossei** 1, 2, and 3 adduct digit 2 to the ulnar side and 4 and 5 to the radial side (i.e., they move the digits toward digit 3, the axis of the hand), while the **dorsal interossei** 1, 2, 3, and 4 abduct digit 2 and 3 to the radial side, and 3 and 4 to the ulnar side (i.e., they move the digits away from the axis of the hand) (Plate 4.16). As explained above, all the intrinsic muscles of the hand are anterior (ventral) muscles, so the name "dorsal interossei" can be confusing. To remember the distinction, keep in mind that the dorsal interossei are innervated by the ulnar nerve, not by the radial nerve that innervates the true posterior (dorsal) muscles of the upper limb.

BOX 4.12 SURPRISING FINDINGS FROM THE STUDY OF HUMAN LIMB PATHOLOGIES AND COMPARATIVE ANATOMY

Some of the authors of this book have recently performed studies (in wild-type nonhuman tetrapods and in humans with congenital defects) focused on a long-standing evolutionary question: Do tetrapods share a general, predictable spatial correlation between limb bones and muscles? We found a surprisingly consistent pattern, both in the **nonpentadactyl limbs** of wild-type taxa such as frogs, salamanders, and chickens, and in the humans with limb birth defects: *The identity and attachments of the distal upper limb and lower limb muscles are mainly related to the physical (topological) position, and not the number of the primordium or even the homeotic identity of the digits to which the muscles are attached.* This type of hard–soft tissue interaction was designated a "nearest neighbor" interaction because the muscle attaches to the nearest neighbor if its normal insertion is not present. Figure 4.6a–d provides numerous examples of "nearest neighbor" interactions in wild-type nonpentadactyl tetrapods. With the loss of digit 1 and/or 5 in the foot or hand, muscles that normally insert on these digits in pentadactyl taxa insert instead on digit(s) 2 and/or 4, respectively. For instance, the hand (manus) digits of urodeles (salamanders) such as axolotls derive from the primordia, and have a homeotic identity, of digits 1–4. However, as shown in Figure 4.6a, the abductor digiti minimi muscle that inserts on digit 5 in pentadactyl tetrapods inserts on digit 4 in salamanders.

These observations are consistent with evolutionary and biomechanical theory in that conserved developmental mechanisms make it possible to preserve functions even when gross morphology changes. The two digits in the most preaxial and postaxial positions (digits 1 and 5, respectively) are often specialized anatomically, and thus may have increased mobility and/or be served by particular muscles, such as the abductor pollicis longus and the abductor digiti minimi. Therefore, loss of one of these digits—for example, digit 1 in the forelimb of birds—may be accompanied by a homeotic transformation (change in segment identity; see Box 6.1) in which the most radial digit of the wing (derived from the primordium of digit 2) receives the identity of digit 1. Even in cases of digit reduction that involve no such homeotic transformations, the "nearest neighbor" mechanism ensures that the extremities retain the muscles and thus the specialized functions of digits 1 and/or 5.

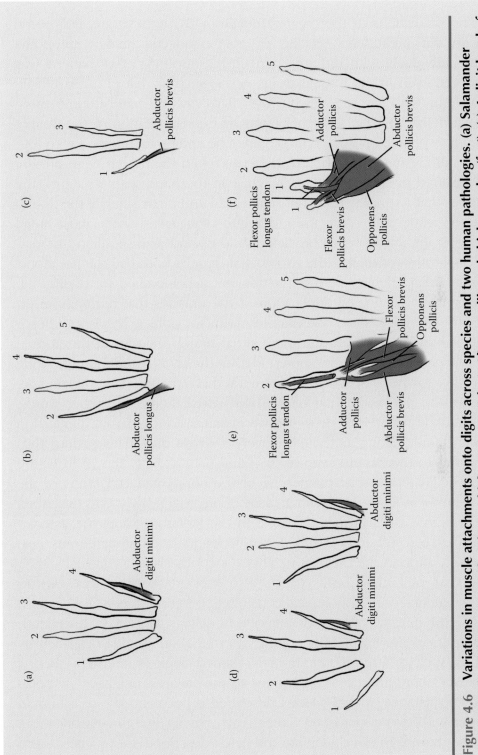

Figure 4.6 Variations in muscle attachments onto digits across species and two human pathologies. (a) Salamander manus (hand); (b) frog manus (hand); (c) chicken manus (wing); (d) crocodile and chicken pedes (feet); (e) 4-digit hand of human trisomy 18 individual; (f) 6-digit hand of human trisomy 18 individual.

The predictions of the "nearest neighbor" hypothesis are also borne out in our preliminary studies of nonpentadactyl human autopodia from individuals with different genetic backgrounds. This pattern was observed even in the most extreme cases, such as that of a nearly full-term fetus with trisomy-18 with four digits on one hand (thumb missing) and six digits on the other hand (partial duplication of the thumb, known as preaxial polydactyly) (Figure 4.6e and f). In the hand with four digits, all the muscles normally associated with the thumb were present but inserted instead onto digit 2, while in the hand with six digits, the muscles that normally insert onto the radial and ulnar sides of the thumb inserted respectively onto the radial and ulnar sides of the most radial and most ulnar thumbs, as predicted. In the hand with six digits, the partial duplication of digit 1 (the two partial thumbs both had the homeotic, or segmental, identity of digit 1) was not accompanied by duplication of metacarpal 1, nor by duplication of the muscles that normally insert onto the thumb. Instead, the muscles abductor pollicis brevis and flexor pollicis brevis, which are the most radial thumb muscles in the normal phenotype, inserted on the more radial of the two partial thumbs, while the adductor pollicis, which is normally the most ulnar thumb muscle, inserted on the more ulnar of the two partial thumbs. In other words, each muscle inserted as though it were "blind" to the partial duplication of the thumb, acting as though it were a single digit. This pattern thus conforms to the pattern seen in the hand with 4 digits shown in Figure 4.6e and in the normal phenotype of the nonpentadactyl limbs of the taxa shown in Figure 4.6f. Other cases of preaxial polydactyly, described in the scarce medical data available in the literature about the associations between soft and hard tissues in nonpentadactyl human limbs, also generally support the "nearest neighbor" hypothesis.

Studies of human birth defects also support the idea that, due to developmental or morphological constraints, a logical, predictable pattern exists even in cases of extreme anatomical defects. Moreover, this pattern often mirrors the normal phenotype of other taxa, an idea defended by authors such as Pere Alberch, in his ill-named "logic of monsters" theory. However, there are a few exceptions to the "nearest neighbor" hypothesis. For instance, in 1957, Heiss described a peculiar case of a human subject with two pentadactyl hands (each having five fingers but missing the thumb) in which, unlike the cases described above, there were apparently no major topological changes to the remaining muscles. That is, in both hands, the normal thumb muscles were all missing. This configuration seems to be characteristic of the rare human disorder named "**tri-phalangeal thumb**," which is a malformation of digit I, including a perfect homeotic transformation of the thumb into an index finger and in which the muscles that are normally associated with the thumb are absent (e.g., abductor opponens/adductor pollicis). If future studies confirm this report, "tri-phalangeal thumbs" will constitute an example of a disorder in which the identity and attachments of the muscles are mainly related to the homeotic identity of the digits to which they attach, and not to their topological position as predicted by the "nearest neighbor" hypothesis. This exception would have important implications: It would demonstrate hitherto-unrecognized developmental plasticity in the patterning of hard and soft tissues, and therefore would open new lines of research within the fields of developmental biology and medicine.

Chapter 5

Lower Limb

The lower limb, also called the pelvic limb, is divided into the **pelvis**, **thigh**, **leg** (**shank**), and **foot** (**pes**). Details about each of these anatomical regions are given in the sections below.

5.1 Lower Limb Skeletal System

5.1.1 Pelvis

The human **pelvis** and associated soft tissues have several unique characteristics not shared by other mammals, including other primates (Plate 5.2). For instance, in humans, the pelvis provides strong support for the vertical vertebral column, and the **gluteal muscles** are anatomically highly modified and are functionally critical to propulsion and stability during **bipedal walking**. Also, adult humans—particularly women, who have a greater **subpubic angle** than men—have relatively larger **pelvic outlets** than other adult primates, allowing enough space for infants with wide shoulders and large brains to pass through the **birth canal**.

The pelvis is formed by two **hip bones (os coxae)** connected posteriorly (dorsally) by the sacrum (which is part of the axial skeleton: See Section 6.1). Each hip bone is made up of three fused bones—the **ilium**, **ischium**, and **pubis**—that meet in the **acetabulum**. The ilium is the most superior bone, and the **anterior superior iliac spine** (at the anteriormost point of the **iliac crest**) is the bony protuberance you feel under the skin when you palpate the "hip bone." The **gluteal lines**, **iliac fossa**, and **arcuate line of ilium** are muscle attachments. In the anterior midline, the **pubic crests** of the left and right pubes are connected via the cartilaginous **pubic symphysis**. The **superior pubic ramus** meets the **ischiopubic ramus** inferiorly, surrounding the oval **obturator foramen**, which is covered by the **obturator membrane**. Other bony features of the pelvis include: **iliopubic eminence, pecten pubis, pubic arch, posterior superior iliac spine**, and **pubic tubercle**. The **anterior sacroiliac ligament** and **posterior sacroiliac ligament** reinforce the **sacroiliac joint**. As their names indicate, the **sacrotuberous ligament** and the **sacrospinous ligament** connect the sacrum to the **ischial tuberosity** and the **ischial spine**, respectively, and thus enclose the **lesser sciatic notch** and the **greater sciatic notch** to form the **lesser sciatic foramen** and the **greater sciatic foramen**, respectively. The latter two foramina are critical landmarks for the muscular and neurovascular structures of the pelvic region.

5.1.2 Hip and Thigh

In the **hip joint**, the acetabulum of the hip bone, enlarged by the cartilaginous **acetabular labrum**, forms the socket for the **head of the femur**, which includes the **fovea for the ligament of the femoral head** (Plate 5.3). This delicate ligament attaches to the **acetabular notch** (spanned by the **transverse acetabular ligament**) and encloses the **artery of the ligament of the head of the femur** (a small branch of the obturator artery). The hip joint allows the femur (and thus thigh) to flex, extend, abduct, adduct, medially rotate, and laterally rotate. The ligaments that form the fibrous joint capsule are the **iliofemoral ligament** (which prevents overextension of the hip joint), **ischiofemoral ligament** (which limits extension of the hip joint), and **pubofemoral ligament**. Inside this capsule against the **lunate surface** of the acetabulum (named for its crescent shape) lies the cartilage-covered **articular surface of the head of the femur**. The head of the femur is connected to the shaft of the femur by the **femoral neck**. Distal to the femoral neck lie the **greater trochanter, lesser trochanter, intertrochanteric line, trochanteric fossa, gluteal tuberosity**, and **pectineal line**, all of which are muscle attachment sites (Plate 5.4). On the distal 3rd of the femur lie additional muscle attachments: the **lateral epicondyle, lateral supracondylar ridge, medial epicondyle, medial supracondylar ridge, lateral lip of linea aspera**, and **adductor tubercle** for the **adductor magnus** muscle. The **lateral condyle** and **medial condyle** articulate with the tibia and are separated by the **intercondylar fossa**, superior to which lies the **popliteal surface**.

5.1.3 Knee and Leg

Bony features for muscle attachment on the **tibial shaft (body)** include, proximally, the **tibial tuberosity** and **soleal line**, and, more distally, the **tibial anterior border** and **medial malleolus** (Plates 5.4 and 5.8). Proximally the fibula includes the **fibular apex, fibular head**, and **fibular neck**, and more distally includes the **fibular shaft (body)**, and then the **lateral malleolus**. The **knee joint** (Plate 5.13a) between the thigh and the leg consists of an articulation between the femur and tibia (specifically, the **superior articular surface, medial condyle, lateral condyle**, and **inter-condylar eminence**; Plate 5.8) and an articulation between the femur and the **articular surface** of the **patella** (the other surface of the patella is the **anterior surface**). This joint mainly permits flexion and extension of the leg, but also permits—and this is often not stated in textbooks and atlases—some lateral (external) and medial (internal) rotation of the leg. Numerous ligaments and tendons are associated with this joint (Plate 5.13a). The **tibial collateral ligament** is attached to the **medial meniscus** through the **joint capsule**. The **fibular collateral ligament** is neither attached to the external surface of the joint capsule nor to the **lateral meniscus** but runs directly from the femur to the fibula. The **posterior cruciate ligament** prevents the tibia from being pushed posteriorly during flexion of the leg and the **anterior cruciate ligament** prevents the tibia from being pulled anteriorly during flexion of the leg; the latter ligament also prevents hyperexten-sion of the leg. Both cruciate ligaments are located inside the joint capsule but outside the **synovial cavity**. The **patellar ligament** connects the tibia and the patella, and is often also designated—incorrectly—as "patellar ten-don." The tendon of the quadriceps femoris muscle complex (which inserts on the patella) has patellar retinacula to help keep the patella centered on the knee. When the leg is fully extended, the knee joint is locked in its most stable position. Lateral rotation of the thigh relative to the leg is needed to unlock the knee joint; this rotation is performed by the **popliteus** muscle when the leg is planted on the ground.

5.1.4 Ankle and Foot

The foot has seven tarsal bones: the **talus** (including the **trochlea**), **cuboid, calcaneus** (including **calcaneal tuberosity** and **sustentaculum tali**), **navicular**, and the **medial (1st), intermediate (2nd), and lateral (3rd) cuneiforms**. It also has five metatarsals (including the distinctive tuberosity of metatarsal 5) and 14 phalanges (Plate 5.14). The **ankle joint** (Plate 5.13b,c) includes the **ankle joint proper** (or **talocrural joint**), which mainly permits dorsiflexion and plantar flexion of the foot, and the **subtalar joint** and **transverse tarsal joint** (which is in turn formed by the **calcaneocuboid joint** and the **talonavicular joint**), which mainly permit **inversion** (movement of the plantar surface toward the body midline) and **eversion** (movement of the plantar surface away from the body midline) of the foot. Several ligaments span these joints: The **medial (deltoid) ligament of the ankle** is composed of the **posterior tibiotalar ligament**, the **tibiocalcaneal ligament**, the **tibionavicular ligament**, and the **anterior tibiotalar ligament**. The lateral ligament of the ankle is composed of the posterior **talofibular ligament**, **calcaneofibular ligament**, and the **anterior talofibular ligament**. The **plantar calcaneonavicular (spring) ligament** supports, together with the **tibialis posterior** tendon, the **head of the talus** and the **longitudinal arch of the foot**. Other significant ligaments of the foot are the **long plantar ligament** and the **short plantar ligament**.

BOX 5.1   ACHONDROPLASIA

Chondrodysplasias are a diverse set of conditions resulting from disruptions of **endochondral bone** development, either during the early stages of cartilage growth and remodeling or later during cartilage-to-bone transition and bone growth. In humans, the most frequent of these conditions is **achondroplasia**, commonly called **dwarfism**, which occurs in 1:20,000 births. Individuals are characterized by short stature with disproportionately shortened upper and lower limbs. The legs are often bowed due to uneven disruption of growth of the **tibia** and **fibula**. Digits are similarly shortened and often the fingers are spread unusually wide apart.

Achondroplasia is an **autosomal dominant condition** (meaning that heterozygous individuals are affected) involving mutations in the **fibroblast growth factor** receptor-3, FGFR3. More than 80% of affected individuals are born to parents of normal stature, making this *the most frequently known gene mutation in humans*. The mutation most often arises during **spermatogenesis**.

5.2 Lower Limb Neurovascular System

5.2.1 Lower Limb Nerves

> **BOX 5.2   EASY WAYS TO REMEMBER: CUTANEOUS INNERVATION OF LOWER LIMB**
>
> Cutaneous and motor innervation of the lower limb is mainly provided by nerves arising from the **lumbosacral plexus**. This plexus includes the **lumbar plexus** (divisions of L1–L4 nerves plus contributions from the sub-costal nerve arising from T12), the **sacral plexus** (S1–S4 nerves, with contributions from the lumbosacral trunk [L4–L5]), and the **pudendal plexus** (including sacral spinal nerves and formerly known as the **coccygeal plexus**). In a simplified way, the mnemonic "3-2-1-1" describes the number of major cutaneous nerves in each part of the lower limb: the anterior thigh (3 nerves), anterior leg (2), posterior thigh (1), and posterior leg (1).
>
> *Three* nerves provide cutaneous innervation of the *anterior (ventral) side of the thigh* (Plate 5.1a). The **lateral femoral cutaneous nerve** arises directly from the **lumbosacral lumbar plexus** and as its name indicates, it mainly innervates the skin ("cutaneous" = of the skin) on the *lateral* region of the anterior side of the thigh (*femoral* region). As also indicated by their name, the **anterior cutaneous branches of the femoral nerve** arise from the **femoral nerve** and innervate the skin on the central (middle) region of the anterior side of the thigh. The **cutaneous branches of the obturator nerve** obviously arise from the **obturator nerve** and innervate the skin of the medial region of the anterior side of the thigh, as well as the medial side of the thigh.
>
> Following the mnemonic "3-2-1-1," *two* nerves provide cutaneous innervation of the *anterior (ventral) side of the leg*. The **medial crural cutaneous nerves** arise from the **saphenous nerve** (which branches from the femoral nerve) and logically innervate the skin on the *medial* region of the anterior side of the leg. The skin of the lateral region of the anterior side of the leg is mainly innervated by the **superficial fibular (peroneal) nerve**, although the more proximal portion of this region also receives contribution from branches—namely, the **lateral sural cutaneous nerve**—of the **sural complex of nerves** that is mainly related to the innervation of the skin of the posterior side of the leg.

Thus, the "1-1" of the mnemonic "3-2-1-1" refers to the fact that the posterior side of the leg is mainly innervated by this sural complex of nerves—which apart from the lateral sural cutaneous nerve also includes the **medial sural cutaneous nerve** and the **sural nerve**—while the posterior side of the thigh is mainly innervated by the **posterior femoral cutaneous nerve** (or **posterior cutaneous nerve of the thigh**) (Plate 5.1b). Note that the lateral sural cutaneous nerve arises, logically, from the **common fibular nerve**, which is more *lateral* than the tibial nerve. Accordingly, the medial cutaneous sural nerve arises from the tibial nerve; distally it meets with a branch of the lateral sural cutaneous nerve—the **sural communicating branch**—to form the sural nerve.

The skin of the gluteal region is innervated by the **cluneal nerves**: The **inferior cluneal nerves** are branches of the **posterior femoral cutaneous nerve**, while the **superior cluneal nerves** and **middle cluneal nerves** are branches of the dorsal rami (Plate 5.1b). Apart from these cutaneous nerves, there are two major nerves in the gluteal region that provide motor innervation to muscles: The **superior gluteal nerve** runs superior to the **piriformis** muscle and the **inferior gluteal nerve** runs inferior to this muscle (Plate 5.7). Also inferior to the piriformis runs—together with the **sciatic nerve** and the posterior femoral cutaneous nerve—the **pudendal nerve**. The pudendal nerve is peculiar because it emerges from the **greater sciatic foramen**, wraps around the sacrospinous ligament, and then turns medially again to pass through the **lesser sciatic foramen** to enter the perineum together with the **internal pudendal artery**. As an aside, the name "pudendal" derives from the Latin word "puta," which refers to the "shameful" supply/drainage of the anal and genital tissues by these vessels and nerves. These classical human anatomists either had a major hang-up about their sexuality or a great sense of humor.

Knowledge of the cutaneous innervation of the lower limb, specifically the cutaneous nerves that arise from the major nerves and which also provide motor innervation of muscles of the lower limb, makes it relatively easy to understand the overall configuration of the major motor nerves of the lower limb (Figure 5.1; Plate 5.5). The single major nerve that provides motor innervation for all the muscles of the anterior compartment of the thigh is the **femoral nerve**, which as stated above (Box 5.2) also gives rise to the **anterior cutaneous branches of the femoral nerve** and to the **saphenous nerve**. The saphenous nerve is easy to find during dissection

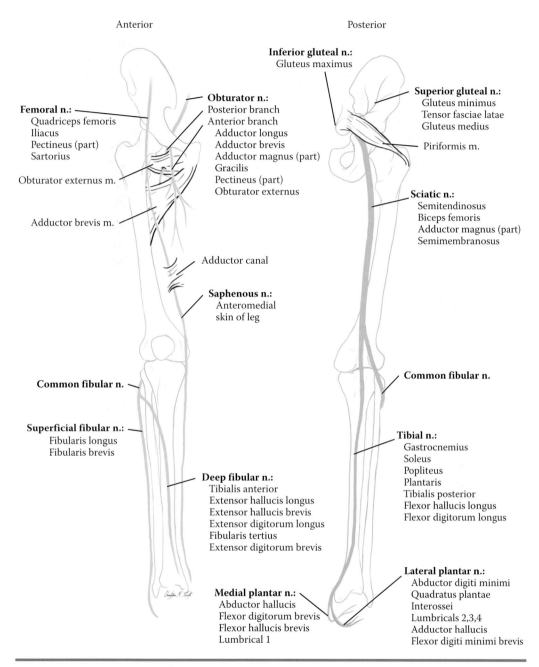

Anterior   Posterior

Inferior gluteal n.:
Gluteus maximus

Superior gluteal n.:
Gluteus minimus
Tensor fasciae latae
Gluteus medius

Obturator n.:
Posterior branch
Anterior branch
Adductor longus
Adductor brevis
Adductor magnus (part)
Gracilis
Pectineus (part)
Obturator externus

Piriformis m.

Femoral n.:
Quadriceps femoris
Iliacus
Pectineus (part)
Sartorius

Obturator externus m.

Adductor brevis m.

Sciatic n.:
Semitendinosus
Biceps femoris
Adductor magnus (part)
Semimembranosus

Adductor canal

Saphenous n.:
Anteromedial
skin of leg

Common fibular n.

Common fibular n.

Superficial fibular n.:
Fibularis longus
Fibularis brevis

Tibial n.:
Gastrocnemius
Soleus
Popliteus
Plantaris
Tibialis posterior
Flexor hallucis longus
Flexor digitorum longus

Deep fibular n.:
Tibialis anterior
Extensor hallucis longus
Extensor hallucis brevis
Extensor digitorum longus
Fibularis tertius
Extensor digitorum brevis

Lateral plantar n.:
Abductor digiti minimi
Quadratus plantae
Interossei
Lumbricals 2,3,4
Adductor hallucis
Flexor digiti minimi brevis

Medial plantar n.:
Abductor hallucis
Flexor digitorum brevis
Flexor hallucis brevis
Lumbrical 1

Figure 5.1   Diagram of nerves of the lower limb, anterior and posterior views.

because it accompanies the femoral artery and vein into the adductor canal region of the thigh, and then the **great saphenous vein** into the leg. The muscles of the medial compartment of the thigh are innervated by the obturator nerve. The **anterior branch of the obturator nerve** and the **posterior branch of the obturator nerve** are separated by the **adductor**

brevis muscle. Also noted in Box 5.2, the **obturator nerve** gives rise to the **cutaneous branches of the obturator nerve**, which innervate the skin of the medial thigh.

Therefore, all the leg and foot muscles as well as the posterior thigh are innervated by nerves coming from the major nerve lying on the posterior side of the thigh: the **sciatic nerve** (Figure 5.1). At the level of the thigh, this major nerve is already divided into the **common fibular division of the sciatic nerve** and **tibial division of the sciatic nerve**, which innervate all the posterior muscles of the thigh. A clear separation between the **common fibular nerve** and the **tibial nerve** is however usually only seen in the distal region of the thigh. As its name indicates, the **common fibular nerve**—which as noted above gives rise to the **lateral sural cutaneous nerve**—divides into two nerves: the **superficial fibular nerve** and the **deep fibular nerve** (Plate 5.9). The superficial fibular nerve innervates the lateral muscles of the leg and part of the skin of the antero-lateral side of the leg as well as the skin of the dorsum of the foot and sends **dorsal digital nerves** to the skin of the toes. The **deep fibular nerve** is the body's "trick" to overcome the problem of not having an anterior nerve, such as the femoral nerve or its branches in the thigh, to innervate the anterior muscles of the leg. The deep fibular nerve thus runs obliquely deep in the anterior compartment of the leg, and then runs deep to the short extensors of the foot. It gives rise to the **dorsal digital branches of the deep fibular nerve** that innervate the skin between the 1st and 2nd toes. The **tibial nerve** passes through the popliteal fossa to run on the posterior side of the leg (Plate 5.10). As noted above (Box 5.2), it gives rise to the **medial sural cutaneous nerve** and thus contributes to the formation of the **sural nerve**, and it innervates all the posterior muscles of the leg. At the calcaneal region, the tibial nerve divides into the **lateral plantar nerve** and the **medial plantar nerve**, which innervate all the intrinsic muscles of the foot and give rise to the **common and proper plantar digital nerves** (Plate 5.16).

5.2.2 Lower Limb Blood Vessels

The **great saphenous vein** is the most prominent of the **superficial veins of the lower limb** (Plate 5.1). This vein arises from the medial end of the **dorsal venous arch of the foot**, passing anteriorly to the **medial malleolus** at the ankle, over the posterior border of the **medial epicondyle of the femur** at the knee, and deeply through the **saphenous opening** to drain into the **femoral vein**. More distally, the great saphenous vein is connected to the deep venous system of the lower limb via several **perforating veins**. More proximally, the great saphenous vein is joined by the **superficial external pudendal vein**, the **superficial epigastric vein**, and the **superficial circumflex iliac vein**. An **accessory saphenous vein** often drains the skin and superficial fascia of the medial side of the thigh. The **small saphenous vein** arises from the lateral end of the **dorsal venous arch** and then runs proximally until it pierces the **deep fascia** in the **popliteal fossa** to join the **popliteal vein**.

The easiest way to learn the deep blood vessels of the lower limb is to start with **the femoral vein** and **femoral artery**, which lie proximally in the anterior side of the **thigh** and form the major supply of blood to, and drainage from, the lower limb (Plate 5.6; Figure 5.2). The **femoral sheath** envelops these two major blood vessels, and has three compartments: The **lateral compartment of the femoral sheath** contains the femoral artery; the **intermediate compartment of the femoral sheath** lies medially to the femoral artery and contains the femoral vein; the **medial compartment of the femoral sheath** lies medially to the femoral vein, contains the femoral lymphatics, and is also called the **femoral canal**, and its proximal opening is called the **femoral ring**. The femoral sheath plus the more lateral **femoral nerve** are located in the **femoral triangle**, which is bounded superiorly by the **inguinal ligament**, laterally by the medial border of the **sartorius** muscle and medially by the **adductor longus** muscle (Plate 5.6a). Just distal to the inguinal ligament, the femoral artery gives rise to three small branches: the **superficial external pudendal artery**, the **superficial epigastric artery**, and the **superficial circumflex iliac artery**. The names of these three small arteries correspond to the names of the three small veins related to the femoral vein (via the great saphenous vein) listed in the previous paragraph. There are only two major arteries and veins in the gluteal region (i.e., the proximal region of the posterior side of the lower limb): The **superior gluteal artery and vein** arise proximally ("superiorly") to the **piriformis** muscle, and the **inferior gluteal artery and vein** arise distally ("inferiorly") to this muscle (Plate 5.7).

There are some similarities between the three major branches of the femoral artery within the femoral triangle (Plate 5.6b) and the three major branches of the axillary and brachial arteries at the level of the proximal arm. The anterior and posterior humeral circumflex arteries and the deep humeral arteries of the upper limb clearly correspond to the **lateral circumflex femoral artery**, **medial circumflex femoral artery**, and the **deep femoral artery (profunda femoris artery** or **deep artery of the thigh)**, respectively. However, in the upper limb, the anterior and posterior

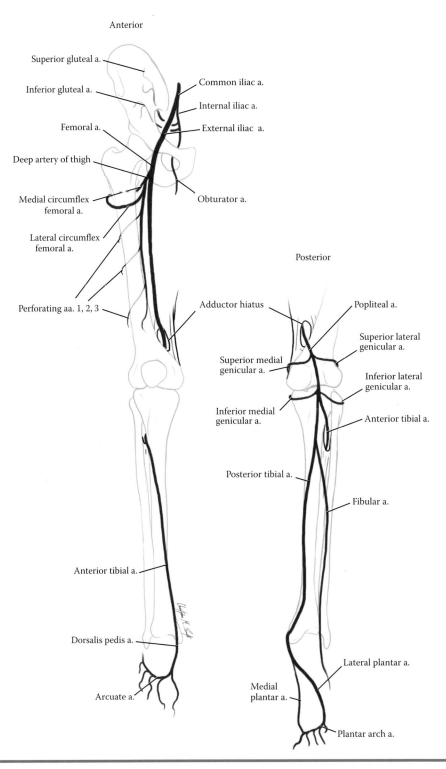

Figure 5.2 Diagram of arteries of the lower limb, anterior and posterior views.

humeral circumflex arteries branch from the axillary artery and the deep humeral artery branches more distally from the brachial artery. In contrast, in the lower limb, all the three analogous structures usually arise close together from the femoral artery, although there are anatomical variations within the femoral triangle.

The femoral artery and its three branches are all located in the anterior compartment of the thigh, but which vessels supply the posterior thigh? The deep femoral artery gives rise to **the perforating branches of the deep femoral artery**, which pierce the **adductor magnus** muscle to provide supply to the medial and posterior thigh muscles. Similar to the "trick" of the deep humeral artery that supplies the posterior compartment of the upper limb, the "trick" of the perforating branches gets around the fact that there is only a major artery entering the arm and thigh, and that this artery (brachial artery in the arm and femoral artery in the thigh) is located on the anterior side.

In contrast, the details of the "trick" performed within the lower limb to provide blood supply to the posterior leg structures are very different from those seen in the "trick" displayed within the upper limb to provide blood supply to the posterior forearm muscles. In the upper limb, an anterior artery—the ulnar artery that branches from the brachial artery—gives rise to the common interosseous artery, which then divides into the anterior interosseous artery and the posterior interosseous artery, the latter of which provides blood supply to the posterior forearm. In the lower limb, the femoral artery runs through the **adductor canal** to reach and pass through the **adductor hiatus**—an opening in the tendon of the adductor magnus muscle just above the knee (Plate 5.6a)—to appear in the posterior side of the knee region. It changes name to **popliteal artery** as it enters the popliteal fossa (Plate 5.6c). As an aside, note that the **saphenous nerve** and the **nerve to the vastus medialis** accompany the femoral artery and femoral vein in the adductor canal, but as noted in Section 5.2.1, the saphenous nerve does not go all the way to the posterior side of the knee region as the popliteal artery and **popliteal vein** do. Instead, the saphenous nerve remains mainly an anterior nerve also in the leg region, to provide cutaneous innervation to the medial side of the leg (Plate 5.1).

As their names indicate, the popliteal artery and vein pass through the **popliteal fossa**, which lies on the posterior side of the knee region and is bordered superolaterally by the **biceps femoris** muscle, superomedially by the **semimembranosus** and the **semitendinosus** muscles, inferolaterally and inferomedially by the two heads of the **gastrocnemius** muscle, posteriorly by skin and deep fascia, and anteriorly by the popliteal surface of the femur and the **popliteus** muscle (Plate 5.7). At this region of the knee, the **popliteal artery** gives rise to branches that form **arterial anastomoses** around the knee joint: the **superior and inferior lateral genicular arteries** and the **superior and inferior medial genicular arteries** ("genu" means "knee" in Latin).

After giving rise to these branches, the popliteal artery runs as the single major artery in the proximal region of the leg. So now, there is the opposite problem that occurs with the femoral artery at the thigh: The femoral artery lies *anteriorly* in the thigh, and therefore it has to do a "trick"—giving rise to the deep femoral artery—to supply the *posterior* side of the thigh; in contrast, the popliteal artery lies posteriorly on the leg, and thus it has to do a "trick" to go to the anterior side of the leg to give blood supply to the anterior leg muscles and surrounding tissues. The "trick" is a bifurcation of the popliteal artery in the proximal leg, giving rise to the **anterior tibial artery** (which pierces the interosseous membrane to enter the anterior side of the leg: Plate 5.9) and to the **posterior tibial artery** (which continues on the posterior side of the leg: Plate 5.11). The posterior tibial artery subsequently—more distally—gives rise to the **fibular artery**. The configuration of the blood vessels of the leg is thus remarkably different from that seen in the forearm. As can be seen in Figures 5.1 and 5.2 and Plates 5.9 and 5.11, the anterior tibial artery runs together with the **deep fibular nerve** in the center of the anterior compartment of the leg, while the posterior tibial artery runs together with the **tibial nerve** through the center of the posterior compartment of the leg. The fibular artery runs mainly alone in the lateral region of the posterior compartment of the leg, whereas the **superficial fibular nerve** runs mainly alone through the lateral compartment of the leg.

The anterior tibial artery crosses the ankle joint, and its name changes to **dorsalis pedis artery** that supplies the dorsum of the foot. It gives rise to the **arcuate artery** that in turn gives rise to the **dorsal metatarsal arteries** (Plate 5.9). The dorsalis pedis artery also gives rise to the **lateral tarsal artery**—which joins the lateral end of the arcuate artery to complete an arterial arch—and to the **deep plantar artery**—which passes between the 1st and 2nd metatarsal bones to enter the sole of the foot to form an anastomosis with the **deep plantar arch** formed mainly by the **lateral plantar artery** (Plate 5.16). The lateral plantar artery arises, together with the **medial plantar artery**, from the **posterior tibial artery**, and gives rise to the **deep plantar arch** that in turn gives rise to **common and proper plantar digital arteries**. Lastly, the **fibular artery** runs on the posterior side of the leg to give rise to the **perforating branch of the fibular**

artery, which pierces the interosseus membrane just above the ankle joint to anastomose with the anterior tibial artery (Plate 5.11).

5.3 Lower Limb Muscular System

To understand the lower limb muscles, one needs to know the difference between the **superficial fascia of the lower limb** and the **deep fascia of the lower limb**. The deep fascia of the limb includes the **fascia lata** in the thigh (the lateral portion of this fascia is particularly strong and is named the **iliotibial tract**), the **crural fascia** in the leg, and the **pedal fascia** in the foot. One also needs to understand that because the pelvic girdle is much less mobile than the pectoral girdle, the muscles normally listed in textbooks and atlases as lower limb muscles are exclusively muscles that attach directly onto the femur, which are **abaxial muscles** (or **appendicular muscles of the pelvic girdle**). Such books often do not include among lower limb muscles those that connect the axial skeleton to the pelvic girdle, such as the **quadratus lumborum** or the **psoas minor**, which are—developmentally—mainly **primaxial muscles** (or **axial muscles of the pelvic girdle**). In contrast, the muscles included as upper limb muscles comprise both primaxial (e.g. serratus anterior, rhomboids, levator scapulae) and abaxial muscles (for more details and for a description of abaxial and primaxial muscles, see Section 4.3).

5.3.1 Muscles of Gluteal Region

The muscles of the gluteal region (Table 5.1) comprise two main developmental units: the ***medial rotator gluteal group*** and the ***lateral rotator gluteal group*** (Plate 5.7). The medial rotator group includes the **tensor fasciae latae**, **gluteus minimus**, and **gluteus medius**, all innervated by the **superior gluteal nerve**. Their common function is to medially rotate the thigh because they attach mainly on the anterior side of the greater trochanter of the femur. The gluteus medius and gluteus minimus are more oblique (fibers running distally and laterally to attach onto the anterior, superior aspect of the greater trochanter of the proximal femur), so they also abduct the thigh because their fibers pass superior to the hip joint (Figure 5.3). The **tensor fasciae latae** is more vertical (fibers running mainly distally to attach through the iliotibial tract to the lateral, proximal tibia), that is more parallel to the femur, so it cannot abduct but has a proper configuration to

Table 5.1 Evolutionary/Developmental Groups of Muscles of Gluteal Region

Group	Muscle	Nerve(s)	Function(s) in Common	Other Function(s)	Artery/ Arteries	Origin	Insertion
Medial rotator gluteal group (posterior pelvic muscles)	Tensor fasciae lata	Superior gluteal n.	Medial rotation of femur	Flexion of femur and tension of fascia lata	Mainly lateral circumflex femoral a. and superior gluteal a.	Iliac crest	Mainly to iliotibial tract
	Gluteus minimus			Abduction of femur	Superior gluteal a.	Ilium	Greater trochanter of femur
	Gluteus medius						
Gluteus maximus (posterior pelvic muscle; functionally part of lateral gluteal rotator group)	Gluteus maximus	Inferior gluteal n.	—	Lateral rotation and extension of femur (proximal and distal parts of muscle also contribute to abduction and adduction of femur, respectively)	Superior and inferior gluteal arteries	Ilium, lumbar fascia, sacrum, sacrotuberous ligament	Gluteal tuberosity of femur and iliotibial tract
Piriformis (posterior pelvic muscle; functionally part of lateral gluteal rotator group)	Piriformis	N. to piriformis (lumbosacral plexus)	—	Lateral rotation and abduction of flexed femur	Superior and inferior gluteal and lateral sacral arteries	Sacrum	Greater trochanter of femur

(Continued)

Table 5.1 (*Continued*) Evolutionary/Developmental Groups of Muscles of Gluteal Region

Group	Muscle	Nerve(s)	Function(s) in Common	Other Function(s)	Artery/ Arteries	Origin	Insertion
Ischiotrochanteric group (functionally part of lateral gluteal rotator group)	Superior gemellus	N. to obturator internus (lumbosacral plexus)	Lateral rotation of femur	–	Inferior gluteal a.	Spine of ischium	Obturator internus tendon
	Obturator internus			–		Ischiopubic ramus and obturator membrane	Greater trochanter of femur
	Inferior gemellus	N. to quadratus femoris (lumbosacral plexus)		–		Ischial tuberosity	Obturator internus tendon
	Quadratus femoris			Lateral rotation on of thigh		Ischial tuberosity	Intertrochanteric crest of femur

Note: All are abaxial muscles.

a. = artery; n. = nerve.

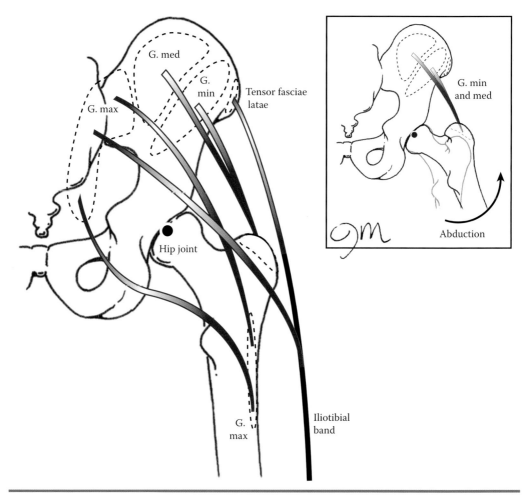

Figure 5.3 **Actions of gluteal muscles: Gluteus maximus (G. max), gluteus medius (G. med), gluteus minimus (G. min), tensor fasciae latae (TFL). Dashed lines indicate areas of attachment and red lines indicate muscle lines of action. Iliotibial band (ITB) and hip center of rotation (CoR) also shown. Inset shows the abduction of the hip by contraction of the gluteus medius and minimus.**

flex the thigh. This muscle can also tense the **fascia lata** (namely, the **iliotibial tract**, as its name suggests, and thus can stabilize the knee joint).

The lateral rotator gluteal group includes the gluteus maximus, innervated by the **inferior gluteal nerve** and the **piriformis, superior gemellus, obturator internus, inferior gemellus**, and **quadratus femoris**, all innervated from nerves from the **lumbosacral plexus**. Apart from lateral rotation, the gluteus maximus also extends the thigh. In addition, the more proximal and more distal fibers of the gluteus maximus can also help abduct and adduct the thigh, respectively (because their fibers pass superior and inferior, respectively, to the hip joint; Figure 5.3). Therefore, when the whole gluteus maximus is contracted, the adduction and abduction components mainly cancel each other, and the muscle mainly contributes to the lateral rotation and extension of the thigh, overall. All five remaining muscles mainly contribute to the lateral rotation of the thigh. The piriformis and quadratus femoris in addition contribute to the abduction and adduction of the thigh, respectively (because their fibers pass superior and inferior, respectively, to the hip joint). Evolutionarily speaking, the piriformis is part of the **posterior pelvic muscles** that also include the gluteus minimus, medius, and maximus and the tensor fasciae lata. The superior gemellus, obturator internus, inferior gemellus, and quadratus femoris, in contrast, are evolutionarily derived from the **ischiotrochanteric pelvic muscles** (Table 5.1).

5.3.2 Muscles of Thigh

The muscles of the thigh are divided into an anterior or extensor compartment of the thigh, a medial or adductor compartment of the thigh, and a posterior or flexor compartment of the thigh (Table 5.2). As explained in Section 5.1, anatomists gave the name "extensor compartment" to the anterior thigh muscles because these muscles extend the leg. However, in the human gross anatomical nomenclature followed by most textbooks and atlases, anterior movement of the thigh (as opposed to the leg) is called "flexion" rather than, more logically, "extension." Although we use the standard anatomical terminology here, from an evolutionary and developmental perspective, it makes more sense to consider anterior movements of the thigh and leg to be flexion, and the posterior movements to be extension. This perspective also makes it easier to identify similarities between the upper and lower limbs (for example, that the quadriceps femoris of the thigh corresponds to the triceps brachii of the arm). Therefore you, as a student, should learn, for your education, the exams, and the communication with your peers, that an anterior movement of the thigh is officially designated as a "flexion of the thigh," but you need to keep in mind that in reality the anterior thigh muscles are developmentally dorsal (extensor) muscles, in order to have a better, deeper knowledge of the lower limb and in particular of the human body as a whole.

The anterior (developmentally dorsal, or extensor) compartment of the thigh includes six muscles: the iliopsoas, the sartorius, and the four muscles that form the quadriceps femoris: the rectus femoris, vastus lateralis, vastus intermedius, and vastus medialis (Plate 5.5a; Table 5.2). The iliopsoas is a

Table 5.2 Evolutionary/Developmental Groups of Muscles of the Thigh

Group	Muscle	Nerve(s)	Function(s) in Common	Other Function(s)	Artery/Arteries	Origin	Insertion
Posterior compartment	Biceps femoris	Sciatic n. (all tibial branch except for short biceps head which is common fibular branch)	Extension of femur and flexion of leg	–	Inferior gluteal, popliteal and perforating arteries	Ischial tuberosity (long head) and femur (short head)	Head of fibula
	Semitendinosus				Inferior gluteal and perforating arteries	Ischial tuberosity	Medial surface of proximal tibia
	Semimembranosus				Inferior gluteal and perforating arteries		
Anterior compartment	Iliopsoas (fused iliacus and psoas major muscles)	Femoral n. (iliacus) and also directly from lumbar plexus (psoas major)	Flexion of femur	–	Medial femoral circumflex a. and iliolumbar a.	Ilium and lumbar vertebrae	Lesser trochanter of femur

(Continued)

Table 5.2 (*Continued*) Evolutionary/Developmental Groups of Muscles of the Thigh

Group	Muscle	Nerve(s)	Function(s) in Common	Other Function(s)	Artery/Arteries	Origin	Insertion
	Sartorius	Femoral n.		Lateral rotation of femur and flexion of leg	Femoral a.	Ilium	Proximal medial tibia
	Rectus femoris		Extension of leg	Flexion of femur	Lateral circumflex femoral a.	Mainly ilium	Patella and then tibial tuberosity via the patellar ligament
	Vastus lateralis			–		Greater trochanter, intertrochanteric line and linea aspera of femur	
	Vastus intermedius				Femoral a.	Lateral and anterior surfaces of femur	
	Vastus medialis					Medial surface of femur	

(Continued)

Table 5.2 (*Continued*) Evolutionary/Developmental Groups of Muscles of the Thigh

Group	Muscle	Nerve(s)	Function(s) in Common	Other Function(s)	Artery/Arteries	Origin	Insertion
Medial compartment	Pectineus	Obturator and femoral nerves	Adduction of femur	Flexion of femur	Obturator a.	Pectineal line of pubis	Pectineal line of femur
	Adductor magnus	Obturator and sciatic (tibial branch) nerves		Extension of femur	Deep femoral a.	Pubis (adductor head, or adductor minimus) and ischial tuberosity (hamstring head)	Linea aspera and adductor tubercle of femur
	Adductor longus	Obturator n.		—		Pubis	Linea aspera of femur
	Adductor brevis			—			Lesser trochanter and linea aspera of femur
	Gracilis			Flexion of leg	Medial circumflex femoral a.	Ischiopubic ramus	Proximal tibia
	Obturator externus			Lateral rotation of femur	Obturator a.	Obturator foramen and membrane	Trochanteric fossa of femur

Note: All are abaxial muscles.

a. = artery; n. = nerve.

composite muscle that mainly flexes the thigh and is formed by two fused muscles: the iliacus (mainly innervated by the femoral nerve) and the psoas major (mainly innervated directly by lumbar nerves of the lumbar plexus). The four muscles that make up the quadriceps femoris (*quad* = four) extend the leg. The rectus femoris is also able to flex the thigh because it originates from the pelvis and not from the femur as do the vastus lateralis, intermedius, and lateralis. The name "sartorius" comes from the Latin word *sartor*, which means tailor, and the sartorius accordingly is often called the tailor's muscle. Its name indicates its three main functions—from anatomical position, you need to perform three movements to sit in the cross-legged posture of early tailors: First you flex the thigh, then you flex the leg, and lastly you laterally rotate the thigh (but not abduct the thigh, as is sometimes stated in anatomical texts). Rather than memorizing the functions of the sartorius, you can logically deduce them: The muscle lies mainly anterior, not lateral, to the femur, so it mainly flexes—not abducts—the thigh. Its fibers are directed distomedially, in an appropriate direction to laterally rotate the thigh. Lastly, its fibers pass posterior to the knee joint, giving it the ability to flex the leg.

BOX 5.3   EASY WAYS TO REMEMBER: MUSCLES OF THE MEDIAL OR ADDUCTOR COMPARTMENT OF THE THIGH

Like the anterior compartment of the thigh, the medial or adductor compartment of the thigh includes six muscles: the obturator externus, pectineus, adductor longus, adductor brevis, adductor magnus, and gracilis (Plate 5.5b; Table 5.2). All these muscles are derived evolutionarily and developmentally from the adductor group of pelvic/thigh muscles, with exception of the pectineus, which seems to be an intermediate muscle receiving contributions from two developmental anlages (primordia): those that give rise to the muscles of the medial and anterior adult thigh. An easy way to learn these muscles is to remember: "*2 additional nerves, 4 additional functions.*" All six muscles are innervated by the obturator nerve and contribute mostly to the adduction of the thigh, and in addition two muscles receive innervation from other nerves and four muscles perform other functions. Logically, the pectineus is one of the muscles that receives additional innervation, namely, from the femoral nerve, because as stated above, it partially derives from the developmental primordium that gives rise to the anterior muscles of the adult thigh (which, of course, are innervated by the femoral nerve). Also due to its association with the anterior thigh muscles, the pectineus has an additional function: to flex the thigh. The other nerve exception is the adductor magnus, which has two heads: the adductor (medial) head (often named "adductor minimus"), and the hamstring (posterior) head, which lies mainly on the posterior side of the thigh. The latter head, as might be expected based on its association with posterior thigh muscles, is innervated by the tibial branch of the sciatic nerve and contributes to extension of the thigh. The two other functional exceptions are: the gracilis, because it inserts onto the posterior and medial side of the proximal leg, can also flex the leg; and the obturator externus, because its fibers pass posterior to the hip joint to insert onto the medial portion of the greater trochanter together with muscles such as the obturator internus, also acts to laterally rotate the thigh. The adductor brevis and adductor longus therefore are essentially adductors and are exclusively innervated by the obturator nerve. The anterior branch of the obturator nerve lies between these two muscles, its posterior branch lying posteriorly to the adductor brevis. The space between the distal portion of the adductor magnus and the femur is called the adductor hiatus. This is the gap through which the femoral vein and artery pass when they go from the anterior side of the thigh to posterior side of the thigh, before changing their names to popliteal vein and artery.

The posterior or flexor compartment of the thigh has only three muscles, all of which act to extend the thigh and flex the leg: the **biceps femoris**, the **semitendinosus**, and the **semimembranosus** (Plate 5.7; Table 5.2). Together, these three muscles are often referred to as the hamstrings. The easy way to tell these muscles apart is to remember that the biceps femoris logically has two heads (*bi* = two; *cephalicus* = head), and inserts on the *lateral* side of the proximal leg. The **short head of the biceps femoris** originates from the lateral portion of the posterior side of the femur, and therefore is innervated by the common fibular branch of the sciatic nerve. The **long head of the biceps femoris** originates from the ischial tuberosity together with the semitendinosus (the *distal* portion of this muscle has a long, thin *tendon*) and the semimembranosus (the *proximal* portion of this muscle has a flat broad *membrane*-like tendon). Therefore, the semitendinosus and semimembranosus—which insert on the *medial* side of the proximal leg—and the biceps long head are logically innervated by the tibial branch of the sciatic nerve.

5.3.3 Muscles of Leg

The muscles of the leg are divided into an **anterior or extensor compartment of the leg**, a **lateral or fibular compartment of the leg**, and a **posterior or flexor compartment of the leg** (Table 5.3). Developmentally and evolutionarily, the anterior and lateral compartments are both derived from an ancestral/common extensor group of muscles. This derivation explains why in adult humans both muscles of the anterior and of the lateral compartments of the leg are innervated by the **common fibular nerve** (anterior compartment by the **deep fibular nerve**; lateral compartment by the **superficial fibular nerve**). The short extensors that insert on the digits (**extensor hallucis brevis** and **extensor digitorum brevis**) are part of the anterior compartment of the leg, and not true intrinsic (ventral, or plantar) foot muscles; accordingly, they are also innervated by the deep fibular nerve. Thus, knowing the developmental histories of the leg muscles makes it easier to understand their innervation patterns without resorting to memorization.

Two fibular muscles make up the **lateral compartment of the leg**: the **fibularis longus** more laterally and **fibularis brevis** more medially (Plate 5.12). As these muscles pass lateral to the ankle joint, they logically evert the foot, and although they derive evolutionarily and developmentally from an ancestral/common extensor group, as noted above, they actually flex (plantarflex) the foot. As its name indicates, the fibularis longus is longer than the fibularis brevis. It originates more proximally from the fibula than the fibularis brevis—and distally—it passes with the brevis deep to the **superior and inferior lateral retinacula** and then continues all the way through the plantar surface of the foot to insert onto metatarsal 1 and the medial cuneiform (the fibularis brevis inserts onto metatarsal 5). Therefore, apart from everting and flexing (plantarflexing) the foot, the fibularis longus also supports the **plantar arches of the foot**.

Continuing from lateral to medial but moving to the **anterior compartment of the leg** (Plate 5.9), we see the **extensor digitorum longus**, which as its name indicates inserts onto digits 2, 3, 4, and 5 to *extend* (dorsiflex) the foot and these *digits*. The tendon to digit 5 often gives rise to a thin tendon called **fibularis tertius**, which inserts on metatarsal 5. The fibularis tertius usually does not have a defined fleshy head, and is a variant structure in the human population—it has been reported to be present in 100% of the population of some geographic/ethnic groups and in only 12% of other groups. Because of these features, anatomists have often stated

Table 5.3 Evolutionary/Developmental Groups of Muscles of the Leg

Group	Muscle	Nerve(s)	Function(s) in Common	Other Function(s)	Artery/ Arteries	Origin	Insertion
Lateral compartment	Fibularis longus	Superficial fibular n.	Eversion and flexion (plantarflexion) of foot	Support of arches of foot	Fibular a.	Proximal fibula	Metatarsal 1 and medial cuneiform
	Fibularis longus			–		Shaft of fibula	Metatarsal 5
Anterior compartment (developmentally closely related to the lateral compartment)	Fibularis tertius	Deep fibular n.	Extension (dorsiflexion) of foot	Eversion of foot	Anterior tibial a.	Usually from tendon of extensor digitorum longus to digit 5	Metatarsal 5
	Extensor digitorum longus			Extension (dorsiflexion) of digits 2, 3, 4, and 5		Proximal tibia and fibula and interosseous membrane	Middle and distal phalanges of digits 2–5
	Extensor hallucis longus			Extension (dorsiflexion) of digit 1 (big toe)		Fibula and interosseous membrane	Distal phalanx of digit 1 (big toe)
	Tibialis anterior			Inversion of foot		Tibia	Metatarsal 1 and medial cuneiform
	Extensor digitorum brevis		–	Extension (dorsiflexion) of digits 2, 3, and 4	Dorsalis pedis a.	Calcaneus	Middle phalanges of digits 2–4
	Extensor hallucis brevis			Extension (dorsiflexion) of digit 1 (big toe)			Proximal phalanx of digit 1 (big toe)

(Continued)

Table 5.3 (*Continued*) Evolutionary/Developmental Groups of Muscles of the Leg

Group	Muscle	Nerve(s)	Function(s) in Common	Other Function(s)	Artery/ Arteries	Origin	Insertion
Posterior compartment	Popliteus	Tibial nerve	–	Flexion and medial rotation of leg (or, if leg is stable, lateral rotation of thigh to unlock the fully looked knee)	Popliteal a.	Postero-lateral surface of distal femur	Medial surface of proximal tibia
	Gastroc-nemius		Plantar flexion (true flexion) of foot	Flexion of leg	Sural arteries	Lateral (lateral head) and medial (medial head) condyles of femur	Calcaneus bone via calcaneal (Achilles) tendon
	Plantaris					Lateral supracondylar ridge of femur	
	Soleus			–		Proximal fibula and tibia	
	Flexor digitorum longus			Plantar flexion (true flexion) of digits 2, 3, 4, and 5	Posterior tibial a.	Shaft of tibia	Distal phalanges of digits 2–5
	Flexor hallucis longus			Plantar flexion (true flexion) of digit 1 (big toe)	Fibular a.	Fibula	Distal phalanx of digit 1 (big toe)
	Tibialis posterior			Inversion of foot	Posterior tibial a.	Tibia and fibula	Navicular and medial cuneiform

Note: All are abaxial muscles.

a. = artery; n. = nerve.

that this muscle is an example of a **vestigial structure**, like the palmaris longus of the upper limb (see Box 4.10). However, the evolutionary history of the two muscles is exactly opposite. The palmaris longus is present in 100% of the studied cases within most mammalian, including primate, taxa, while in humans, its frequency is *lower*, approximately 85% in many groups. The fibularis tertius is instead absent in all studied mammalian taxa except humans. Therefore, the muscle probably was recently acquired in the human lineage and might well be linked to efficient **terrestrial bipedalism**.

Still proceeding from lateral to medial, medially to the extensor digitorum longus muscle lies the muscle that inserts on digit 1 (big toe), the **extensor hallucis longus**. Lastly, and logically, the medial of the **long extensors of the leg** is the **tibialis anterior**, which inserts onto the medial side of the foot and so extends (dorsiflexes) and inverts the foot. Deep to the long extensors of the leg lie the two deeper and shorter muscles of the anterior compartment: the **extensor hallucis brevis** that logically *extends* (dorsiflexes) digit 1 (*hallux* or big toe) and the **extensor digitorum brevis** that logically *extends* (dorsiflexes) the other *digits*. Unlike all the other muscles named "digitorum" in both the upper and the lower limb, the extensor digitorum brevis inserts on digits 2–4, instead of digits 2–5.

The muscles of the **posterior compartment of the leg** are divided into a more proximal/superficial layer (Plate 5.10) with four muscles—**popliteus, gastrocnemius, plantaris**, and **soleus**—and a more distal/deep layer with three muscles (Plate 5.11)—**flexor digitorum longus, flexor hallucis longus**, and **tibialis posterior**. All these muscles are evolutionarily and developmentally derived from an ancestral/common flexor group of muscles, so all are innervated by the tibial nerve and all except the popliteus can plantarflex the foot (plantar flexion is "true" flexion in the evolutionary/developmental sense, as opposed to dorsiflexion, which would be considered extension). The **popliteus** does not plantarflex the foot because it does not cross the ankle, instead running from the lateral surface of the posterior side of the distal femur to the medial surface of the posterior side of the proximal tibia, thus logically being a flexor and a medial rotator of the leg. However, the popliteus is one of the few muscles for which students need to know *not only the action on the bone on which the muscle inserts but also on the bone from which the muscle originates*. This is because the popliteus plays a crucial role in unlocking the fully extended knee by laterally rotating the femur when the leg is fixed. See Section 5.1 for more details about how lateral rotation of the femur helps to unlock the knee when the leg is fully extended.

The **plantaris**, **soleus**, and the two-headed **gastrocnemius** lie mainly superficial (posterior) to the popliteus. These three muscles are closely related evolutionarily and developmentally and all insert together onto the calcaneus bone of the foot through the **calcaneal tendon** (or **Achilles tendon**). Because the very thin plantaris and the massive gastrocnemius originate from the femur and thus cross the **knee joint** in addition to the **ankle joint**, both these muscles also contribute to the flexion of the leg. A strikingly similar pattern is found in the upper limb, where the thin palmaris longus—which in terms of **topology** (**anatomical configuration**) corresponds to the plantaris of the lower limb—also contributes to the flexion of the forearm (remember that in the human anatomical position used in most current atlases, the anterior surface of the upper limb corresponds to the posterior surface of the lower limb; see Section 2.3 and Figure 2.3).

BOX 5.4   EASY WAYS TO REMEMBER: DISTAL/DEEP MUSCLES OF THE POSTERIOR COMPARTMENT OF THE LEG

The **distal/deep muscles of the posterior compartment of the leg** (Plate 5.11) are some of the most confusing because their configuration is so different from that of the anterior leg muscles. The common mnemonic "**T**om-**D**ick-**H**arry" can be useful, but only if it is used properly. This mnemonic describes the order of the tendons from medial to lateral (first **t**ibialis posterior, then flexor **d**igitorum longus, and then flexor **h**allucis longus) *exclusively* at the level of the distal leg/ankle region. At this level, these tendons pass deep (anterior) to the **superior and inferior flexor retinacula** and posterior to the medial malleolus. However, the mnemonic does not apply to the more proximal region of the leg, where the major portions of the muscles lie. For example, in the proximal leg, the **flexor digitorum longus** is actually the most medial of the three muscles. Therefore, instead of "Tom-Dick-Harry," we suggest the mnemonic *"two-one-zero" steps away from the logical position*. First of all, the "logical position" of each of these muscles based on their insertions (and the corresponding muscles in the anterior compartment), from medial to lateral, would be: (1) tibialis posterior, most medial muscle because it inserts on the medial (tibial) aspect of the ankle as its name indicates; (2) flexor hallucis longus because it inserts on digit 1, which is the most medial digit; and (3) flexor digitorum longus most laterally because it attaches to the most lateral digits of the foot (digits

2–5). The only muscle that is not in its logical position at the levels of the proximal leg is the flexor digitorum longus. The relationship of the two other muscles is always logical, similar to that of the corresponding muscles of the anterior compartment of the leg (the tibialis posterior and the extensor hallucis longus): The **tibialis posterior** is always medial to the **flexor hallucis longus**, as it should logically be. So, students just need to remember that, at the level of the proximal leg, the flexor digitorum longus is *two steps away* from its logical position by being the most medial of the three muscles, while it should logically be the most lateral one. At the level of the distal leg/ankle region, the tendons of the flexor digitorum longus cross the tendon of the tibialis anterior so they now lie between the most medial tendon (tibialis anterior) and the most lateral tendon (flexor hallucis longus), therefore lying *only one step away* from their logical position (because they should be the most lateral tendons, logically). Then, more distally on the plantar surface of the foot, the tendons of the flexor digitorum longus lie *zero steps away* from their logical position because they cross the tendon of the flexor hallucis longus to attach onto the more lateral digits.

The flexor digitorum longus, tibialis anterior, and flexor hallucis longus contribute to plantar flexion (true flexion) of the foot, each of them serving an additional function: The flexor digitorum longus flexes digits 2–5, the tibialis posterior inverts the foot, and the flexor hallucis longus flexes digit 1.

5.3.4 Muscles of Foot

As noted above, drawing topological correspondences between the upper and lower limb can make it easier and quicker to learn their anatomy. This is particularly true for the extrinsic (originating in the leg) and intrinsic structures of the foot (Plates 5.15 through 5.17); especially the muscles, which are strikingly similar to the intrinsic structures of the hand. In fact, in both the hand and foot, there are

1. Superficial short abductors of digits 1 (**abductor hallucis** in the foot; abductor pollicis brevis in the hand) and of digits 5 (abductor digiti minimi in both the foot and the hand) that attach onto the base of the proximal phalanx of these digits.

2. Superficial flexors (**flexor digitorum brevis** on the foot; flexor digitorum superficialis on the hand) that mainly attach onto the middle phalanges of digits 2–5 through tendons that bifurcate distally.
3. Between these bifurcated tendons emerge the tendons of the extrinsic (leg/forearm) deep long flexors (**flexor digitorum longus** on the foot; flexor digitorum profundus on the hand) that insert onto the distal phalanges of digits 2–5.
4. Moreover, from the tendons of these deep long flexors originate, both in the hand and the foot, four small intrinsic muscles—the **lumbricals** 1, 2, 3, and 4—that attach onto both the base of the proximal phalanx and the extensor expansions of digits 2, 3, 4, and 5, respectively.
5. Additionally, there is a separate long extrinsic (leg/forearm) muscle that inserts on the distal phalanx of digit 1 (**flexor hallucis longus** on the foot; flexor pollicis longus on the hand) and that somewhat divides the "medial" and "lateral" heads of the intrinsic muscle **flexor hallucis brevis** of the foot and the "superficial" and "deep" heads of the intrinsic muscle flexor pollicis brevis of the hand, which insert on the base of the proximal phalanx of digit 1 in both the foot and the hand.
6. On the opposite side of the foot and hand—that is, the side of digit 5—an intrinsic muscle **flexor digiti minimi brevis** inserts on the base of the proximal phalanx of digit 5.
7. Deep to the muscles listed above, an intrinsic adductor muscle (**adductor hallucis** on the foot; adductor pollicis on the hand) attaches to the proximal phalanx of digit 1 and is divided into a **transverse head** and an **oblique head**.
8. Between the adductor pollicis/hallucis muscle and the bones of digit 1 lies a short, thin intrinsic muscle that usually inserts onto the base of the proximal phalanx of this digit and is unfortunately not described in atlases and textbooks although it is often (in the foot), or even usually (in the hand), present (**adductor pollicis accessorius** or "**volaris primus of Henle**" in the hand; **adductor hallucis accessorius** or "**interosseous plantaris hallucis**" in the foot).
9. Deep to the adductor pollicis/hallucis, there are three intrinsic ventral interossei (**plantar interossei** in the foot; plantar interossei in the hand) that insert onto the proximal phalanges of, and adduct, the digits, and four intrinsic **dorsal interossei** that insert onto the proximal phalanges of, and abduct, the digits.
10. In addition to this striking similarity between the extrinsic and intrinsic muscles of the hand and foot, the pattern of innervation of all these

muscles is also remarkably similar. All the intrinsic hand muscles listed just above are innervated by the ulnar nerve, and the intrinsic muscles of the foot are innervated by its "equivalent" (**lateral plantar nerve**), with the exception of the short abductors and flexors of digit 1 and some lumbrical muscles, which are innervated by the median nerve in the hand and its "equivalent" in the foot (**medial plantar nerve**).

In fact, there are only a few major differences that the students need absolutely to know between the intrinsic muscles of the foot and of the hand:

1. Unlike the hand, the foot has no opponens muscles for digits 1 and 5. Many primate species have opponens muscles in the foot: For instance, they often have an opponens hallucis because the big toe is highly mobile and widely separated from the other toes. But the opponens hallucis was lost during the evolutionary history of the human lineage, when our big toe moved immediately adjacent to the other toes and lost some of its mobility.

2. Because the big toe is much wider compared to the other toes ("This little piggy went to market…"), the main axis of the foot is digit 2 and not digit 3 as in the hand. Therefore, while the hand **lumbricals** 1 and 2 (going to digits 2 and 3) are innervated by the median nerve, only lumbrical 1 of the foot (going to digit 2) is innervated by the corresponding nerve, the medial plantar nerve.

3. Accordingly, contrary to the hand where the adduction performed by the palmar interossei and abduction by the dorsal interossei is defined relative to digit 3, in the foot adduction by the **plantar interossei** and abduction by the **dorsal interossei** is defined relative to digit 2. Therefore, the three plantar interossei muscles insert on the medial sides of digits 3, 4, and 5 to adduct these toes, while the four dorsal interossei in the foot insert on the medial side of digit 2 and lateral sides of digits 2, 3, and 4, to abduct digit 2 medially and to abduct digits 2, 3, and 4 laterally.

4. In the foot, there is no muscle similar to the intrinsic muscle palmaris brevis of the hand. On the other hand, the foot has a broad muscle **quadratus plantae** that has no equivalent in the hand. The quadratus plantae runs from the calcaneus to insert onto the tendons of the flexor digitorum longus and is innervated by the lateral plantar nerve, that is the nerve that corresponds to the ulnar nerve innervating the palmaris

Table 5.4 Evolutionary/Developmental Groups of Intrinsic Foot Muscles

Group	Muscle(s)	Nerve(s)	Function(s)	Artery/Arteries	Origin	Insertion
Lumbricals	Lumbricals 1, 2, 3, and 4	Medial plantar n. (lumbrical 1) and lateral plantar n. (lumbricals 2, 3, and 4)	Flexion of proximal and extension of middle and distal phalanges of digits 2, 3, 4, and 5, respectively	Medial and lateral plantar arteries	Tendons of flexor digitorum longus	Base of proximal phalanges and extensor expansions of digits 2, 3, 4, and 5 (therefore reaching both the middle and distal phalanges of each digit)
Abductor hallucis and flexor hallucis brevis	Abductor hallucis	Medial plantar n.	Abduction of proximal phalanx of big toe	Medial plantar a.	Calcaneus	Base of proximal phalanx of big toe
	Flexor hallucis brevis (medial and lateral heads)		Flexion of proximal phalanx of big toe	Medial plantar and 1st plantar metatarsal arteries	Cuboid and lateral cuneiform	Proximal phalanx of big toe
Quadratus plantae	Quadratus plantae	Lateral plantar n.	Helping flexor digitorum longus in flexion of distal phalanges of digits 2, 3, 4, and 5	Medial and lateral plantar arteries and plantar arch	Calcaneus	Tendons of flexor digitorum longus
Abductor digiti minimi and flexor digiti minimi brevis	Abductor digiti minimi	Lateral plantar n.	Abduction proximal phalanx of digit 5	Mainly lateral plantar a.	Plantar aponeurosis	Base of proximal phalanx of digit 5
	Flexor digiti minimi brevis		Flexion of proximal phalanx of digit 5		5th metatarsal	Base of proximal phalanx of digit 5

(Continued)

Table 5.4 (Continued) Evolutionary/Developmental Groups of Intrinsic Foot Muscles

Group	Muscle(s)	Nerve(s)	Function(s)	Artery/Arteries	Origin	Insertion
Adductors	Adductor hallucis (transverse and oblique heads)	Lateral plantar n.	Adduction of proximal phalanx of big toe	Mainly lateral plantar a.	Metatarsal 3 (oblique head) and metatarso-phalangeal ligaments and surrounding structures (transverse head)	Proximal phalanx of big toe
	Adductor hallucis accessorius (often present, also named interosseous plantaris hallucis)		Mainly unknown, might be insignificant or, for instance, weak adduction and/or flexion of proximal phalanx of big toe		Often base of metatarsal 1	Base of proximal phalanx of big toe and surrounding structures
Interossei	Dorsal interossei 1, 2, 3, and 4	Lateral plantar n.	Abduct digit 2 to medial side and digits 2, 3, and 4 to lateral side	Lateral tarsal, medial tarsal, plantar and dorsal metatarsal, and dorsal digital arteries	Metatarsals	Medial side of proximal phalanx of digit 2, and lateral side of proximal phalanges of digits 2, 3, and 4
	Plantar interossei 1, 2, and 3		Adduct digits 3, 4, and 5 to medial side			Medial side of proximal phalanx of digits 3, 4, and 5

Note: All originally from the ventral, or flexor, group of muscles) muscles of the foot plus the quadratus plantae muscle, which is actually an extrinsic (leg) muscle (all these muscles are abaxial muscles.

a. = artery; n. = nerve.

brevis in the hand. The quadratus plantae is usually described in atlases and textbooks as an intrinsic foot muscle, and is listed in Table 5.4 accordingly, but it is actually derived evolutionarily from the ancestral flexor group of leg muscles, and by that definition, it is an extrinsic muscle of the foot.

With these simple comparisons between the extrinsic and intrinsic muscles and their innervation, function and/or overall configuration, in the foot and hand, together with the information summarized in Table 5.4, it will hopefully be much more comprehensive and logical (and thus easier) to understand and recall the foot muscles and associated structures.

Chapter 6

Trunk

6.1 Trunk Skeletal System

Some basic concepts of the development of trunk musculoskeletal tissues have already been covered in Section 2.1 and Boxes 4.1 and 4.2. The **axial skeleton** is composed of the skull, ribs, sternum, and vertebral column (Plate 6.2). The vertebral column consists of 7 cervical, 12 thoracic, 5 lumbar, 5 sacral, and 3–5 coccygeal vertebrae. This column of bones forms the axis of the body and protects the spinal cord. **Cervical, thoracic, and lumbar vertebrae** are easily distinguished from one another by several criteria. Generally, vertebrae from more superior regions of the **vertebral column** have more oval-shaped **vertebral bodies** and relatively larger **vertebral foramina**. Each vertebral foramen, through which the spinal cord runs, is bounded on the ventral (anterior) side by the vertebral body and on the dorsal (posterior) side by **vertebral arch**, formed by the **pedicles** and **lamina**. The shapes of the vertebral arches change over a cranio-caudal gradient that reflects their function: The main function of the upper vertebral column is to allow flexibility and movement of the neck and head, while the main function of the lower vertebral column, particularly the sacral region, is to provide support to the upper body and pelvic girdle. In fact, as will be explained below, cervical vertebra 1 (**atlas**) has no body at all but instead encircles the **dens** of cervical vertebra 2 (**axis**), allowing the head to rotate through a large range of motion (Plate 3.31b,c).

Another obvious difference between vertebral regions is that only **thoracic vertebrae** have facets for rib articulations. The rib articulations are located on both the **transverse processes** (**transverse costal facets**, because the Latin for ribs is "costae") and the **vertebral bodies** (**superior**

and inferior costal facets). The superior and inferior costal facets, also called "demi-facets," should not be confused with the **superior and inferior articular processes** of each vertebra, which form synovial joints with the corresponding inferior and superior processes of the adjacent vertebrae. The **superior and inferior vertebral notches** of adjacent vertebrae form an **intervertebral foramen**, for the passage of the **spinal nerves**. In addition to intervertebral contacts at articular processes, the cranial (superior) and caudal (inferior) surfaces of adjacent vertebral bodies articulate through a fibrocartilaginous **intervertebral disc**. As an interesting exercise, measure your height in the morning and evening and, especially if you have been sitting most of the day, you will probably have decreased in height by a few millimeters. This shrinkage is due to normal compression of your intervertebral discs. Another way of distinguishing the cervical, thoracic, and lumbar vertebrae is that the spinous processes of the thoracic vertebrae usually are long, slender, and directed inferiorly; those of the cervical ones are often shorter and bifurcated at the tip; those of the lumbar ones are usually broad and project posteriorly. In addition, the **cervical vertebrae** are the only vertebrae whose transverse processes are perforated by a **foramen transversarium** (Plate 6.3). C1 (**atlas**) is easily recognized by the absence of a body, which during development forms the **dens** of C2 (**axis**) (Box 6.2), and by its **posterior tubercle**, **posterior arch**, **groove for the vertebral artery**, and its articulation with the occipital bone.

The five **sacral vertebrae** fuse to form the **sacrum**, which articulates with the most inferior lumbar vertebra at the **sacral promontory** and **superior articular facet of the sacrum** (Plate 6.4c). The **median sacral crest** corresponds to the fused **spinous processes**. The **medial furrow** is a midline depression on the posterior (dorsal) surface just superficial to the median sacral crest. Unlike other regions of the vertebral column, the sacrum has both **posterior (dorsal) sacral foramina** and **anterior (ventral) sacral foramina**. These foramina do not exactly match each other anatomically (the 1st posterior sacral foramen is not exactly the same size as the 1st anterior sacral foramen, and so on). The **sacral canal** is a continuation of the vertebral canal, which contains the spinal cord in the more superior regions of the body, but the sacral canal is partly open posteriorly; this opening is called the **sacral hiatus**. The three to five fused **coccygeal vertebrae** form the **coccyx**, which is a **rudimentary structure** in humans because we lack tails. In vertebrate animals with tails, the vertebrae located caudal to the sacral vertebrae (usually posterior to the sacrum, because these animals are quadrupeds) are many and often stout, and they are called caudal rather than coccygeal.

BOX 6.1   EVOLUTION OF *HOX* GENES

It has long been known that each somite "knows" its location along the body axis as soon as it is formed. However, the genetic basis for regional specification remained elusive until the discovery and analysis of a family of transcription factors produced by ***Hox* genes**. The hypothesis that individual genes might specify body regions was put forth based on a set of features found in mutagenized *Drosophila* (fruit fly) embryos. In some of these flies, one body segment developed with an anatomical identity characteristic of a different **body segment**. For example, some flies had wings on their 3rd thoracic segment in addition to the normal 2nd thoracic set of wings; the unknown gene whose mutation caused this abnormality was therefore named *bithorax*. In other flies, legs developed in the place of antennae; the responsible gene was called *antennepedia* ("antenna feet"). Little else was known about the mutated genes until methods for gene cloning and sequencing became available. Then a set of eight similar genes, all of which produce segment-identity changes when mutated, were discovered. These were called **homeotic genes**. Each gene has, in the fruit fly, a nearly identical 180-base-pair region called the *homeobox*, and the peptide sequence generated from this region binds to DNA, thus acting as a **transcription factor**. In the fly, all these homeotic genes are located on the same chromosome. Each is expressed early in larval development in a different region of the larval body. The most surprising discovery about these genes is that their alignment along the 3′ to 5′ direction on the chromosome *matches the alignment of their expression along the cranio-caudal axis of the larva.* Nobody anticipated that chromosome anatomy would presage or mirror embryonic anatomy. These genes do not cause segments to form; rather, they impart within each segment (or set of adjacent segments) the information necessary to develop with characteristics appropriate to that part of the body. Collectively, these eight genes are called the ***Hox* family** of transcription factors.

This family of body axis-specifying genes has been retained throughout animal evolution. Early in the evolution of vertebrates, *the entire set of Hox genes underwent several duplications*, and the duplicated families were translocated to other chromosomes. Thus, most vertebrates, including all mammals, have *four* nearly complete sets of *Hox* genes, and the entire set is referred to as the *Hox* gene family. Some members of the *Hox* gene family either failed to duplicate or were lost in evolution, and some new members were added (e.g., *Hox* genes 10, 11, 12, and 13). The higher-numbered *Hox* genes are duplications of lower-numbered *Hox* genes, and most of these are involved in limb development. The expression of many *Hox* genes begins during **gastrulation**, generally sweeping from cranial to caudal within the **neural plate** and **paraxial mesoderm**. While expression can occur over an extended domain, *the expression of most Hox genes has a precise cranial boundary, which is different for each member of a Hox cluster.* Boundaries of *Hox* gene expression in paraxial mesoderm (somites) in a mammal are shown in Plate 6.1a. The first indication that this phenotype–genotype relationship might not be a coincidence came from comparisons of *Hox* gene expression patterns in somites of birds and mammals. Most mammals have seven **cervical vertebrae**, but chickens have 16. In chick embryos, expression boundaries of *Hox* genes in groups 5 and 6 cluster around somites 20–22, which is the future **cervico-thoracic junction**, while in mammals, this clustering occurs at somites 11–13, which also corresponds to the future cervico-thoracic junction.

The **tubercle of a rib** articulates with the transverse costal facet of the thoracic vertebra of the same number. The **head of a rib** articulates with the superior costal facet of the body of that same vertebra, and with the inferior costal facet of the body of the vertebra that lies just above it (Figure 6.1) (the superior and inferior facets are often called demi-facets because they form half of a surface). For example, rib 5 articulates with transverse and superior costal facets of vertebra T5, and with inferior costal facet of vertebra T4. The exceptions to this rule are vertebra T1, which has a demi-facet on the lower portion of its body for the head of rib 2, but has a complete articular facet in the upper portion of its body for the head of rib 1, and vertebrae T9 through T12, which usually articulate with one rib each. Vertebrae T9 and T10 often have a transverse costal facet plus a single costal facet on their bodies to articulate with a single rib, while vertebrae T11 and T12 often lack facets on their transverse processes for articulation with the ribs. Other significant features of most ribs, apart from the head and tubercle, are the **neck, angle, shaft (body)**, and the **costal groove**. The anteromedial end of most ribs is connected to a **costal cartilage**. The costal cartilage is attached directly to the sternum in ribs 1–7 (**true ribs**) and to the costal cartilage of the rib above in ribs 8–10 (**false ribs**). In ribs 11–12

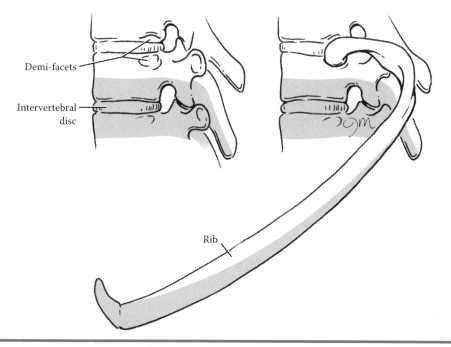

Demi-facets

Intervertebral disc

Rib

Figure 6.1   Contributions of adjacent somites to vertebrae and ribs: shaded areas represent contributions of one somite, and unshaded areas represent contributions from the adjacent somite.

(**floating ribs**), the costal cartilage does not articulate with the skeleton but instead ends in the **abdominal musculature**.

The sternum is made up of the **manubrium, body of the sternum, and xiphoid process**. The superior margin of the manubrium is called the **jugular notch (suprasternal notch)**. Its inferior margin is connected to the body of the sternum at the **sternal angle**. The inferior margin of the body of the sternum articulates with the xiphoid process at the **xiphisternal junction**.

As mentioned above, most thoracic vertebrae (usually T2–T8) articulate with two ribs, forming two demi-facets on the lateral sides of their vertebral body. The two demi-facets only form a complete, circular facet when the inferior costal facet of one vertebra (e.g., T3) is combined with the superior costal facet of the following vertebra (e.g., T4). This configuration is actually easy to understand from a developmental perspective. During development, each **somite** gives rise to the inferior half of one vertebra (including its inferior costal facet), the superior half of the next vertebra (including its superior costal facet and its transverse process and thus transverse costal facet), the intervertebral disc between these halves, and most of the proximal portion of the rib articulating with both these halves (Figure 6.1; see Box 6.2).

BOX 6.2   FORMATION AND OSSIFICATION
OF VERTEBRAE AND RIBS

As **sclerotome** cells spread longitudinally (rostrally and caudally) during development, populations from each somite become contiguous with their neighbors, transiently obscuring their segmental origin. On the contrary, superficial **myotome** and **dermamyotome** populations retain their original segmental separation. At the same time, sclerotomal cell populations expand dorsally and ventrally, to eventually surround the entire **neural tube** and **notochord**. A segmental organization of sclerotomes then reappears, but is *shifted half a segment from its original pattern*. Differential rates of cell proliferation and aggregation establish alternating zones of sclerotome tissue having greater and lesser densities. The *caudal part of each former sclerotome is denser*, and this joins with *the cranial, less dense population* from the adjacent sclerotome to form a single **vertebra**: Thus, *each vertebra is formed from halves of two adjacent sclerotomes* (Figure 6.1).

This reorganization of the original segmental relationships is fundamental to establishing the proper connections among motor nerves, skeletal muscles, and vertebrae, and also critical for the morphogenesis of **intervertebral joints**. Myotomes do not shift position. Thus, after sclerotome resegmentation, each myotome spans two adjacent vertebrae. In this way, muscle contraction causes adjacent vertebrae to move relative to each other. If the resegmentation did not occur, the muscles would connect the cranial and caudal ends of the same vertebra. In some cases, muscles from several adjacent myotomes fuse together, obscuring evidence of their multisegmental origin. However, their original segmental organization is often revealed by the presence of segmental transverse bands of connective tissues. For example, "washboard abs" reflect the multisegmental origin of the rectus abdominis muscle (located in the ventral abdominal body wall).

At the cranial end of the vertebral column, the segmental reorganization of sclerotomal cells deviates from the pattern described above. A mesenchymal aggregate associated with the 5th somite, which based on its position should be the embryonic body of the 1st cervical vertebra, the **atlas**, fuses instead with the cranial aspect of the 2nd cervical vertebra, the **axis**, to form part of its cranial articular surface and the **dens**. Thus, the body, cranial articular surface, and dens of the axis develop from five ossification centers, while the small body of the atlas forms from only one.

In contrast to the multisegmental origin of vertebrae, each axial **myotome** that gives rise to the **back muscles** during development derives from a single somite. The consequence of this developmental process is that the muscles of the back are attached to two successive vertebrae, allowing the vertebral column to bend (both functionally in the adult and during development). The spinal curvatures produced by bending of the vertebral column during development are critically important for our bipedal species because, unlike quadrupedal animals whose center of gravity typically is located near the center of the torso, our center of gravity lies around the center of the pelvis. Our marked **lumbar curvature** and broad lumbar vertebrae thus help to bring the center of gravity closer to the midline and above the feet, while our **cervical curvature**, together with our strong **nuchal ligament** (a structure found in just a few other animals), helps keep our heads upright (Plate 6.4a).

BOX 6.3 EVOLUTION, DEVELOPMENT, AND PATHOLOGY: SEVEN CERVICAL VERTEBRAE, CANCER, CONSTRAINTS, *HOX* GENES

The differences listed above between the different types of vertebrae are the result of strong **evolutionary constraint** and **developmental constraint**, and are therefore particularly important in the context of human medicine and pathology. For instance, the number of lumbar vertebrae is different among mammals, including large apes, humans, and our close fossil relatives. The greater number of human lumbar vertebrae contributes to a more flexible lower back, allowing swiveling of the hips and trunk during bipedal walking. However, almost all mammals have seven cervical vertebrae, no matter how short or long their necks (Figure 6.2). Chickens, on the other hand, have 16 cervical vertebrae, and the number of cervical vertebrae is extremely varied in other reptiles and in amphibians and fish, so this developmental constraint seems to be specific to mammals.

Accordingly, anatomical variations in humans and other mammals almost never include changes in the number of cervical vertebrae. Why is this so? Functionally, it is difficult to imagine how having, for instance, an eighth cervical vertebrae in an adult human (or an adult giraffe) would make it more difficult to survive long enough to

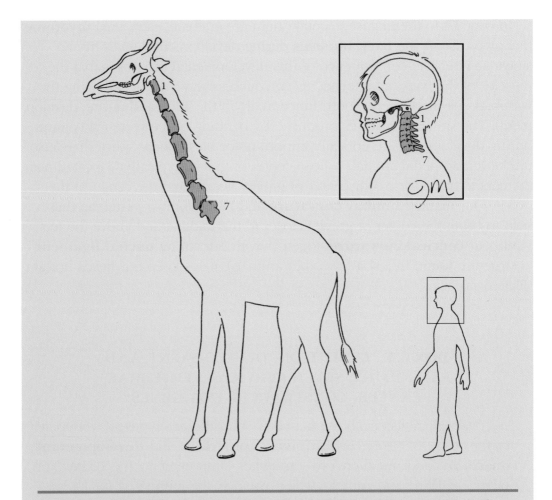

Figure 6.2 **Illustration showing the result of developmental constraints in the number of cervical vertebrae in mammals. All mammals have seven cervical vertebrae, whether they have long necks (e.g., giraffe) or short necks (e.g., humans), although the number of thoracic and lumbar vertebrae can vary greatly.**

reproduce. According to Frietson Galis, a particularly brilliant evolutionary and developmental biologist, the answer seems to be that the changes in *Hox* **gene expression** (Plate 6.1) that could alter cervical vertebrae number during development are also associated with neural problems and an increased susceptibility to early childhood **cancer** and stillbirth. As explained in Box 6.1, *Hox* **genes** are involved in the development of the skeletal axis and nervous system. In mammals such as humans, *Hox* genes also play roles in both normal and abnormal (cancerous) cell proliferation. Several types of cancer in young children are associated with abnormalities in *Hox* gene expression, and in cases of embryonic cancers, the incidence of a cervical rib appears to be increased. Therefore, the problem is not the number of cervical vertebrae in the adult, but that having more or fewer cervical vertebrae is symptomatic of serious disruptions to earlier and very important stages of human development such that the embryo would be unlikely to survive. This is a clear illustration of the importance of developmental constraints, in contrast to **functional constraints**, in the evolutionary history of humans and other mammals, and of why the human body is best understood not simply as an adult functional machine, but instead as an organism derived from complex evolutionary and developmental processes.

6.2 Trunk Neurovascular System

The **spinal nerves** divide into **ventral rami** and **dorsal rami**. The ventral rami form the **intercostal nerves** (T1–T12) in the thoracic region, as well as the **cervical plexus, brachial plexus, lumbar plexus**, and **sacral plexus**, thereby supplying the muscles and skin of the upper and lower limbs and part of the trunk. The lumbar plexus, sacral plexus, and **pudendal plexus** form the **lumbosacral plexus**. Intercostal nerves T1 to T11 lie together with the **intercostal veins** and **intercostal arteries** in the 11 **intercostal spaces**, while T12 is a **subcostal nerve** that courses below the 12th **rib**. The anterior end of an intercostal space is supplied by the **anterior intercostal branches** of the **internal thoracic artery**. The intercostal nerves supply the **serratus posterior superior** and **serratus posterior inferior** and the **external intercostal muscle, internal intercostal muscle**, and **innermost intercostal muscle**. The intercostal nerves also have **cutaneous branches** at the anterior (ventral) and lateral surfaces of the thoracic wall. The dorsal rami supply the **paravertebral muscles (paraspinal muscles)** and **skin** near the midline of the back.

BOX 6.4   BASIC CONCEPTS IN CARDIOVASCULAR EVOLUTION

The basic concepts in cardiovascular development were given in Section 3.3.2 and Box 3.15 (see Plate 3.4). Evolutionarily speaking, the cardiovascular system is a remarkable feature constrained by a long evolutionary history. It meets the continuously changing needs of the embryo/fetus/placenta, and has allowed tetrapods to move from an aquatic, gill-based respiratory system, to a terrestrial, air-breathing environment. Its initial formation follows a highly conserved blueprint that has been operating with few changes since the early vertebrates evolved. This blueprint specifies, for example, such basic features as left-right symmetry, bilateral origins of cardiac tissues, and the formation and then remodeling of segmentally arranged aortic arches. In fact, it is now known that even non-vertebrate animals such as urochordates (also named tunicates; this clade includes the sea squirts) have a heart that shares various features with the heart of vertebrate animals, such as its strong developmental link with branchiomeric muscles (Figure 3.1).

There are special demands on the cardiovascular system that must be met for the rapidly growing embryo to survive. The organism quickly becomes too large to have its needs for oxygen, electrolytes, and metabolic precursors met by diffusion, and a more effective delivery system must be established. This requires the rapid expansion of both intra- and extra-embryonic vascular networks and a pumping mechanism to move blood. Mammalian embryos also develop extensive—and highly specialized—arrangements of vascular networks in the placenta.

During all stages of heart and vascular development, functional circulation must be maintained. This moving column of fluid places exceptional hemodynamic pressures on the continuously expanding and remodeling network of channels and the pump.

Finally, at birth, the cardiovascular system must be able to rapidly switch from placental to internal systems of gas exchange and metabolic regulation in the lungs, liver, kidneys, and intestines, all of which are minimally functional during fetal stages. At any other time of life, rapidly increasing the vascular pressure to these organs would cause massive internal hemorrhaging.

BOX 6.5 DEFECTS RELATED TO THE
DEVELOPMENT OF THE DUCTUS ARTERIOSUS

In Section 3.3.2 and Box 3.15, we described the normal development of the **ductus arteriosus** (see Plate 3.4). Coarctation of the aorta is a condition in which the aorta closes just proximal (upstream) of the site where the ductus normally enters. It has been shown in animal models of coarctation that this site on the aorta develops a ductus-like wall structure.

Patent ductus arteriosus (PDA) is a failure of the ductus to close after birth. The extent of failure is variable; small openings are usually not clinically important. However, larger persistent shunts allow mixing of oxygenated and deoxygenated blood. The direction of flow can be "left to right" or "right to left." These terms refer to the side of the **heart** from which the blood is derived. Blood from the left ventricle, richly oxygenated, leaves the heart via the (left 4th) aortic arch. If the ductus persists, this blood, usually under higher pressure than that leaving the right ventricle, will flow from the aorta through the ductus into the pulmonary stream and return to the **lungs**. This is a left-to-right flow. Over time, this overloading of the right outflow will cause the wall of the **right ventricle** to get stronger, which can cause the flow through a patent ductus to reverse, dumping deoxygenated blood into the aorta, downstream of the branches to the **head**.

One of the few genes thus far linked to isolated PDA is AP2B. This is a **transcription factor** that is normally expressed in **neural crest cells**. Why neural crest cells? Embryology offers an explanation: Crest cells are the source of all smooth muscles associated with aortic arches and also contribute to the wall of the ductus arteriosus, as explained in Chapter 3. AP2B activates genes involved with (1) oxygen monitoring and (2) production of endothelin-1, a potent vasoconstriction enhancer of perivascular smooth muscles. PDA is a significant complicating problem in human infants born more than 8 weeks prematurely, with an increase in frequency of almost 10% for each week in advance of 32 weeks of gestation. Several syndromes of unrelated genetic origin can be accompanied by PDA.

6.3 Trunk Muscular System

6.3.1 Back Muscles

The "**back muscles**" as defined in atlases and textbooks of human anatomy include several subgroups of muscles with completely different evolutionary and developmental origins. One of these subgroups, the "**superficial muscles of the back**" (or "**posterior thoracoappendicular muscles**"), includes a head muscle (the **trapezius**) innervated by a cranial nerve (**CN XI**) and several upper limb muscles (**latissimus dorsi**, **rhomboid major**, **rhomboid minor**, and **levator scapulae**) innervated by **ventral rami**. These muscles are accordingly described in Sections 3.4 and 4.3 with the other neck and upper limb muscles.

Another subgroup of the "back muscles" is the "**intermediate muscles of the back**." This intermediate group is formed by the **serratus posterior superior** and the **serratus posterior inferior** muscles. The serratus posterior superior originates from the **nuchal ligament** and spinous processes of vertebrae C7 to T2, and inserts onto ribs 2–5. The serratus posterior inferior originates from the spinous processes of vertebrae T11 to L2 and inserts onto ribs 8–12. Both muscles pull the ribs cranially, thereby enlarging the thoracic cavity and causing the lungs to expand. These respiratory muscles are innervated by the intercostal nerves formed by the thoracic ventral rami, as noted above.

Therefore, the only group of the so-called "back muscles" that is truly innervated by the dorsal rami of spinal nerves encompasses the "**deep muscles of the back**" of current textbooks and atlases (Plate 6.5). This deep muscle group includes, among others, the splenius capitis and cervicis, the semispinalis capitis, cervicis, and thoracis, the multifidus and the rotatores, and the longissimus, spinalis, and iliocostalis. The **splenius capitis** is easily recognizable because it lies deep to the trapezius. The fibers of the splenius capitis course obliquely across the neck to insert onto the mastoid process of the temporal bone and the superior nuchal line of the occipital bone. The **splenius cervicis** is continuous with the splenius capitis, but attaches onto the transverse processes of vertebrate C1 to C4. Both muscles originate from the nuchal ligament and the spinous processes of vertebrate C7 to T6 and are involved in the extension, rotation, and lateral flexion of the head/neck (Plate 3.31a).

The **erector spinae** group mainly extends the vertebral column when both sides work together, and bends the vertebral column laterally toward the side that is active when only one side is contracted. This group is

> **BOX 6.6   EASY WAYS TO REMEMBER: ATTACHMENTS OF SPINALIS, LONGISSIMUS, AND ILIOCOSTALIS**
>
> One easy way to memorize the most inferior attachments of the three muscles of the erector spinae group is to think "one step further": Inferiorly, the spinalis only reaches the lumbar vertebrae; the longissimus goes further inferiorly/laterally, reaching the sacrum; and the iliocostalis even further, reaching the ilium.

divided into three muscles: The **spinalis** runs from spinous processes to spinous processes, covering the lumbar, thoracic, and cervical regions; the **longissimus** runs from the sacrum to the transverse processes of thoracic and cervical vertebrae, as well as to the mastoid process of the temporal bone via the **longissimus capitis** bundle; and lastly, as its name indicates, the **iliocostalis** runs from the *ilium* (iliac crest) to the ribs (*costae*).

The **transversospinalis group** is deep to the erector spinae group and includes muscles that mainly connect the transverse and spinous processes of the vertebrae. These connections allow rotational and lateral (i.e., transverse) bending movements between adjacent vertebrae, and stabilize the vertebral column. This transversopinalis group includes the semispinalis group, described below: the **rotatores, interspinales** and **intertransversarii,** and the **multifidus,** which runs from the sacrum, ilium crest, and erector spinae aponeurosis to the vertebral spinous processes.

The **semispinalis group** is divided into the **semispinalis capitis** (from transverse processes of the upper thoracic vertebrae to occipital bone), **semispinalis cervicis** (from transverse processes of the upper five or six thoracic vertebrae to spinous processes of C1 to C5) (Plate 3.31a), and **semispinalis thoracis** (or **semispinalis dorsi**; from transverse processes of T6 to T10, to spinous processes of C6 and C7). The semispinalis capitis extends the head and is easily distinguished from the splenius capitis because its fibers course vertically and parallel to the long axis of the neck (not obliquely), and because the **greater occipital nerve** passes through the semispinalis capitis.

6.3.2 Suboccipital Muscles

The **suboccipital muscles** are innervated by dorsal rami and lie deep to the back muscles named the splenius capitis and semispinalis capitis

(Plate 3.31a). They include, among others, the **obliquus capitis inferior muscle** (spinous processes of C2 to transverse processes of C1), the **obliquus capitis superior muscle** (transverse processes of C1 to occipital bone), and the **rectus capitis posterior major muscle** (spinous processes of C2 to occipital bone), which form the boundaries of the **suboccipital triangle**. Medial to these is the **rectus capitis posterior minor**, which extends from the posterior tubercle of the atlas to the occipital bone. These muscles extend and laterally bend the head at the **atlanto-occipital joints**, and rotate the head at the **atlanto-axial joints**. In the suboccipital triangle lie the **vertebral artery** and the **suboccipital nerve** (unique among dorsal primary rami in that it has no cutaneous distribution).

Suggested Readings

For All Chapters

Grant's Atlas of Anatomy. 13th edition, 2012.

Grant's Dissector. 15th edition, 2015.

Gray's Atlas of Anatomy. 2nd edition, 2014.

Netter's Atlas of Human Anatomy. 6th edition, 2014.

Aiello, L. and C. Dean. 1990. *An Introduction to Human Evolutionary Anatomy*. Academic Press (San Diego).

Bardeen, C. R. 1910. Development of the skeleton and of the connective tissues. *Manual Human Embryol*. 1:438–439.

Bergman, R. A., S. A. Thompson, A. K. Afifi, and F. A. Saadeh. 1988. *Compendium of Human Anatomic Variation: Text, Atlas, and World Literature*. Urban & Schwarzenberg (Baltimore, USA).

Boughner, J. and C. Rolian (eds.). 2016. *Developmental Approaches to Human Evolution*. John Wiley & Sons (Hoboken, USA).

Carlson, B. M. 2013. *Human Embryology and Developmental Biology*. Elsevier Saunders (Philadelphia, USA).

De Beer, G. 1940. *Embryos and Ancestors*. Clarendon Press (Oxford, UK).

Diogo, R. and V. Abdala. 2010. *Muscles of Vertebrates: Comparative Anatomy, Evolution, Homologies and Development*. Taylor & Francis (Oxford, UK).

Diogo, R. and B. Wood. 2012. *Comparative Anatomy and Phylogeny of Primate Muscles and Human Evolution*. Taylor & Francis (Oxford, UK).

Diogo, R. and B. Wood. 2012. Violation of Dollo's law: Evidence of muscle reversions in primate phylogeny and their implications for the understanding of the ontogeny, evolution and anatomical variations of modern humans. *Evolution* 66:3267–3276.

Fleagle, J. G. 2013. *Primate Adaptation and Evolution*. 3rd edition. Academic Press (San Diego, USA, Waltham, USA, London, UK).

Gebo, D. L. 2014. *Primate Comparative Anatomy*. Johns Hopkins University Press (Baltimore, USA).

Goodrich, E. S. 1958. *Studies on the Structure and Development of Vertebrates*. Dover Publications (New York, USA).

Held, L. I. 2009. *Quirks of Human Anatomy—An Evo-Devo Look at the Human Body.* Cambridge University Press (Cambridge, USA).

Held, L. I. 2014. *How the Snake Lost its Legs.* Cambridge University Press (Cambridge, USA).

Lewis, W. H. 1910. The development of the muscular system. In: *Manual of Embryology*, Vol 2. (eds. Keibel F, Mall FP), pp. 455–522. JB Lippincott (Philadelphia, USA).

Lieberman, D. E. 2013. *The Story of the Human Body: Evolution, Health and Disease.* Pantheon Press (New York, USA).

Moore, K. L., T. V. N. Persaud, and M. G. Torchia. 2013. *The Developing Human: Clinically Oriented Embryology.* Elsevier/Saunders, Medical (Philadelphia, USA), 540pp.

Schoenwolf, G. C. 2009. *Larsen's Human Embryology.* Elsevier (Churchill Livingstone), Medical (Philadelphia, USA).

Shubin, N. 2008. *Your Inner Fish: A Journey Into the 3.5-Billion-Year History of the Human Body.* Pantheon Books (New York, USA).

Smith, C. M., J. M. Ziermann, J. L. Molnar, M. C. Gondre-Lewis, C. Sandone, E. T. Bersu, M. A. Aziz, and R. Diogo. 2015. *Muscular and Skeletal Anomalies in Human Trisomy in an Evo-Devo Context: Description of a T18 Cyclopic Newborn and Comparison between Edwards (T18), Patau (T13) and Down (T21) Syndromes Using 3-D Imaging and Anatomical Illustrations.* Taylor & Francis (Oxford, UK).

Willis, R. A (ed.). 1962. *The Borderland of Embryology and Pathology (2nd edition).* Butterworths (London, UK).

For Introduction

Abzhanov, A. 2013. Von Baer's law for the ages: Lost and found principles of developmental evolution. *Trends Genet.* 29:712–722.

Alberch, P. 1989. The logic of monsters: Evidence for internal constraint in development and evolution. *Geobios Mém. Spéc.* 12:21–57.

Arthur, W. 1997. *The Origin of Animal Body Plans: A Study in Evolutionary Developmental Biology.* Cambridge University Press (Cambridge, UK).

Blumbergm M. S. 2009. *Freaks of Nature: What Anomalies Tell Us About Development and Evolution.* Oxford University Press (New York, USA).

Carroll, S. B. 2005. *The New Science of Evo-Devo: Endless Forms Most Beautiful.* WW Norton & Company (New York, USA).

Darwin, C. 1859. *On the Origin of Species by Means of Natural Selection, or the Preservation of Favored Races in the Struggle for Life.* John Murray (London, UK).

Dawkins, R. 2004. *The Ancestor's Tale—A Pilgrimage to the Dawn of Life.* Houghton Mifflin (Boston, USA).

Diogo, R., M. Linde-Medina, V. Abdala, and M. A. Ashley-Ross. 2013. New, puzzling insights from comparative myological studies on the old and unsolved forelimb/hindlimb enigma. *Biol. Rev.* 88:196–214.

Gluckman, P. 2009. *Principles of Evolutionary Medicine, 9th Edition*. Oxford University Press (Oxford, UK).

Gould, S. J. 1977. *Ontogeny and Phylogeny*. Harvard University Press (Cambridge, USA).

Gould, S. J. 2002. *The Structure of Evolutionary Theory*. Belknap (Harvard, USA).

Hall, B. K. 1984. Developmental mechanisms underlying the formation of atavisms. *Biol. Rev.* 59:89–124.

Leroi, A. M. 2003. *Mutants: On the Form, Varieties and Errors of the Human Body*. Harper Collins (London, UK).

Ramon y Cajal, S. 1937. *Recollections of My Life*. Translated by E. Horne Craigie. American Philosophical Society (Philadelphia, USA).

Shapiro, B. L., J. Hermann, and J. M. Opitz. 1983. Down syndrome—A disruption of homeostasis. *Am. J. Med. Genet.* 14:241–269.

Singer, C. 1957. *A Short History of Anatomy and Physiology from the Greeks to Harvey*. Dover Publications (New York, USA).

Wood, J. 1867. On human muscular variations and their relation to comparative anatomy. *J. Anat. Physiol.* 1867:44–59.

For Head

Crelin, E. S. 1987. *The Human Vocal Tract: Anatomy, Function, Development, and Evolution*. Vantage Press Inc. (New York, USA).

Czajkowski, M. T., C. Rassek, D. C. Lenhard, D., Bröhl, and C. Birchmeier. 2014. Divergent and conserved roles of Dll1 signaling in development of craniofacial and trunk muscle. *Dev. Biol.* 395:307–316.

de Ruiter, M. C., A. C. Gittenberger-de Groot, S. Rammos, and R. E. Poelmann. 1989. The special status of the pulmonary arch artery in the branchial arch system of the rat. *Anat. Embryol (Berl).* 179:319–325.

Diogo, R., R. Kelly, L. Christiaen, M. Levine, J. M. Ziermann, J. Molnar, D. Noden, and E. Tzahor. 2015. A new heart for a new head in vertebrate cardiopharyngeal evolution. *Nature* 520:466–473.

Edgeworth, F. H. 1935. *The Cranial Muscles of Vertebrates*. Cambridge University Press (Cambridge, USA).

Ericsson, R., R. Knight, and Z. Johanson. 2013. Evolution and development of the vertebrate neck. *J. Anat.* 222:67–78.

Gasser, R. F. 1967. The development of the facial muscles in man. *Am. J. Anat.* 120:357–376.

Gopalakrishnan, S., G. Comai, R. Sambasivan, A. Francou, R. C. Kelly, and S. Tajbakhsh. 2015. A cranial mesoderm origin for esophagus striated muscles. *Dev. Cell.* 34:694–704.

Köntges G. and A. Lumsden. 1996. Rhombencephalic neural crest segmentation is preserved throughout craniofacial ontogeny. *Development* 122:3229–3242.

Lieberman, D. E. 2011. *The Evolution of the Human Head*. The Belknap Press of Harvard University Press (Cambridge, USA).

Noden, D. M. and P. Francis-West. The differentiation and morphogenesis of craniofacial muscles. *Dev. Dyn.* 235:1194–1218.

Wind, J. 1970. *On the Phylogeny and Ontogeny of the Human Larynx.* Wolters-
Noordhoff (Groningen, USA).

For Upper Limb

A good summary of some concepts related to postcranial musculoskeletal development and primaxial versus abaxial muscles, including Q&A concerning
related pathological case studies, is given in https://web.duke.edu/anatomy/
embryology/limb/limb.html.

Bello-Hellegouarch, G., M. A. Aziz, E. M. Ferrero, M. Kern, N. Francis, and
R. Diogo. 2013. "Pollical palmar interosseous muscle" (musculus adductor
pollicis accessorius): Attachments, innervation, variations, phylogeny, and
implications for human evolution and medicine. *J. Morphol.* 274:275–293.

Cihak, R. 1972. Ontogenesis of the skeleton and intrinsic muscles of the human
hand and foot. Adv. Anat. Embryol. *Cell Biol.* 46:1–194.

Diogo, R., S. Walsh, C. Smith, J. M. Ziermann, and V. Abdala. 2015. Towards the
resolution of a long-standing evolutionary question: Muscle identity and
attachments are mainly related to topological position and not to primordium
or homeotic identity of digits. *J. Anat.* 226:523–529.

Durland, J. L., M. Sferlazzo, M. Logan, and A. C. Burke. 2008. Visualizing the lateral
somitic frontier in the Prx1Cre transgenic mouse. *J. Anat.* 212:590–602.

Heiss, H. 1957. Bilateral congenital triphalangeal thumb: Five-fingered hand in
mother and child. *Z. Anat. Entw-Gesch.* 120:226–231.

Lewis, O. J. 1989. *Functional Morphology of the Evolving Hand and Foot.* Clarendon
Press (Oxford, UK).

Matsuoka, T., P. E. Ahlberg, N. Kessaris, P. Iannarelli, U. Dennehy, W. D.
Richardson, A. P. McMahon, and G. Koentges. 2005. Neural crest origins of the
neck and shoulder. *Nature* 436:347–355.

Negus, V. E. 1949. *The Comparative Anatomy and Physiology of the Larynx.* Hafner
Publishing Company (New York, USA).

Shearman, R. M. and A. C. Burke. 2009. The lateral somitic frontier in ontogeny
and phylogeny. *J. Exp. Biol.* 312B:602–613.

Valasek, P., S. Theis, A. DeLaurier, Y. Hinits, G. N. Luke, A. M. Otto, J. Minchin
et al. 2011. Cellular and molecular investigations into the development of the
pectoral girdle. *Dev. Biol.* 357:108–116.

For Lower Limb

Bardeen, C. R. 1906. Development and variation of the nerves and the musculature
of the inferior extremity and of the neighboring regions of the trunk in man.
Am. J. Anat. 6:259–390.

Cihak, R. 1972. Ontogenesis of the skeleton and intrinsic muscles of the human hand and foot. *Adv. Anat. Embryol. Cell Biol.* 46:1–194.

Diogo, R. and J. L. Molnar. 2014. Comparative anatomy, evolution and homologies of the tetrapod hindlimb muscles, comparisons with forelimb muscles, and deconstruction of the forelimb-hindlimb serial homology hypothesis. *Anat. Rec.* 297:1047–1075.

Lewis, O. J. 1989. *Functional Morphology of the Evolving Hand and Foot.* Clarendon Press (Oxford, UK).

For Trunk

Christ, B., R. Huang, and J. Wilting. 2000. The development of the avian vertebral column. *Anat. Embryol.* 202:179–194.

Galis, F. 1999. Why do almost all mammals have seven cervical vertebrae? Developmental constraints, Hox genes, and cancer. *J. Exp. Zool. (Mol. Dev. Evol.)* 285:19–26.

Galis, F. and J. A. J. Metz. 2003. Anti-cancer selection as a source of developmental and evolutionary constraints. *BioEssays* 25:1035–1039.

Plates

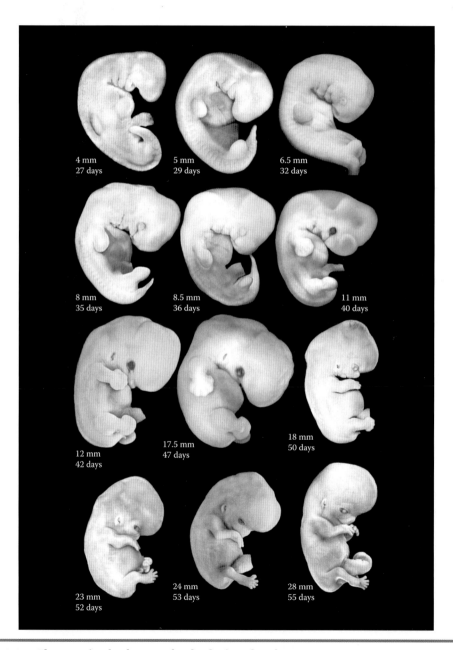

4 mm
27 days

5 mm
29 days

6.5 mm
32 days

8 mm
35 days

8.5 mm
36 days

11 mm
40 days

12 mm
42 days

17.5 mm
47 days

18 mm
50 days

23 mm
52 days

24 mm
53 days

28 mm
55 days

Plate 2.1　Changes in the human body during development.

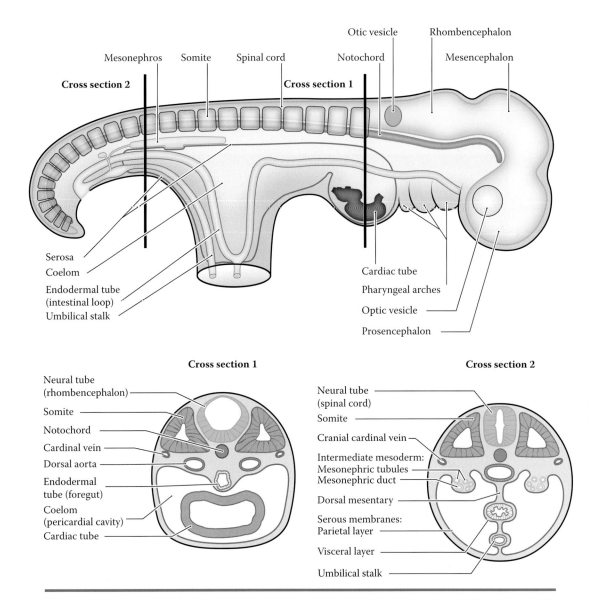

Otic vesicle

Rhombencephalon

Mesonephros   Somite   Spinal cord   Notochord   Mesencephalon

Cross section 2   Cross section 1

Serosa
Coelom
Endodermal tube
(intestinal loop)
Umbilical stalk

Cardiac tube
Pharyngeal arches
Optic vesicle
Prosencephalon

Cross section 1   **Cross section 2**

Neural tube
(rhombencephalon)
Somite
Notochord
Cardinal vein
Dorsal aorta
Endodermal
tube (foregut)
Coelom
(pericardial cavity)
Cardiac tube

Neural tube
(spinal cord)
Somite
Cranial cardinal vein
Intermediate mesoderm:
Mesonephric tubules
Mesonephric duct
Dorsal mesentary
Serous membranes:
Parietal layer
Visceral layer
Umbilical stalk

Plate 2.2   Schematic sketch of a "typical" mammalian embryo, viewed from the right side. Locations of the central nervous system (blue), notochord (red rod), somites (pink rectangles), urogenital structures (green), digestive tract (yellow), and heart are shown. The body is fully enclosed and separated from extra-embryonic membranes except at the umbilical stalk. Cross sections show cuts at two different antero-posterior (cranio-caudal) locations.

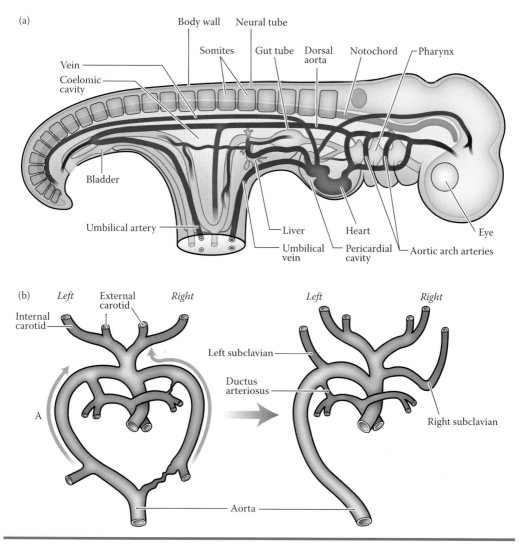

Plate 2.3 **Vertebrate development and aortic arch remodeling. (a) Key anatomical features that define the embryonic body plan of vertebrate embryos. (b) Schematic dorsal views showing the remodeling of the 4th and 6th aortic arches and associated arteries. Note especially the loss on the right side of both the dorsal part of the 6th aortic arch and the dorsal aorta caudal to the subclavian branch.**

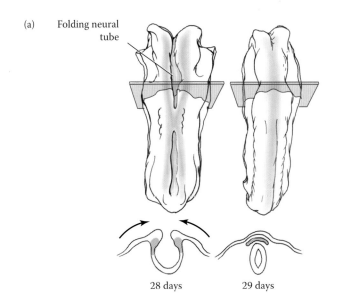

(a) Folding neural tube

28 days   29 days

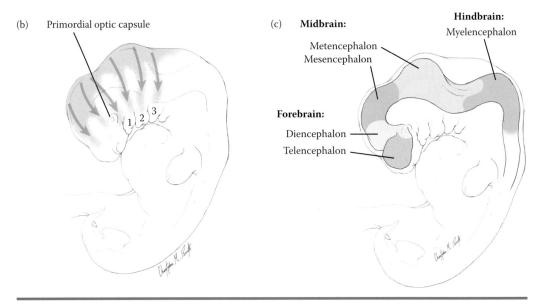

(b) Primordial optic capsule

1 2 3

(c) **Midbrain:**

Metencephalon
Mesencephalon

Hindbrain:
Myelencephalon

Forebrain:

Diencephalon
Telencephalon

Plate 3.1   Neural crest and brain development in human embryos. (a) Neural tube folding in 28- and 29-day-old embryos, dorsal view. Neural crest cells (shown in purple) originate from the "crest" of the folding neural tube and pinch off after neural tube fusion, neural crest cells; (b) cranial neural crest migration in 35-day-old embryo; (c) regions of the developing brain in 35-day-old embryo.

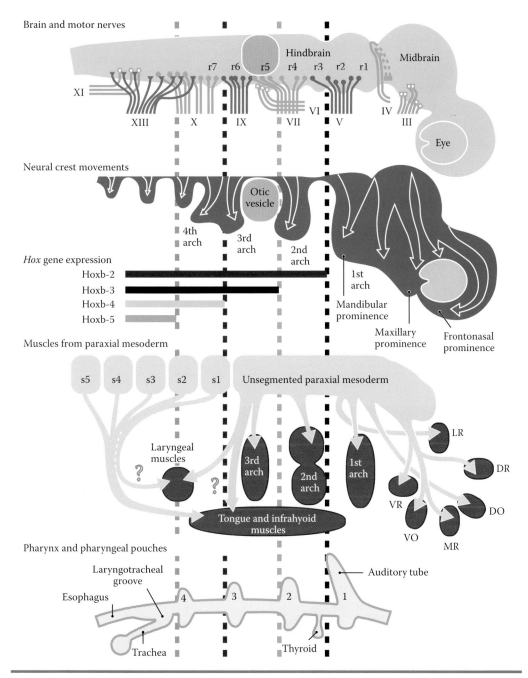

Plate 3.2   Developmental origins of head and neck muscles.

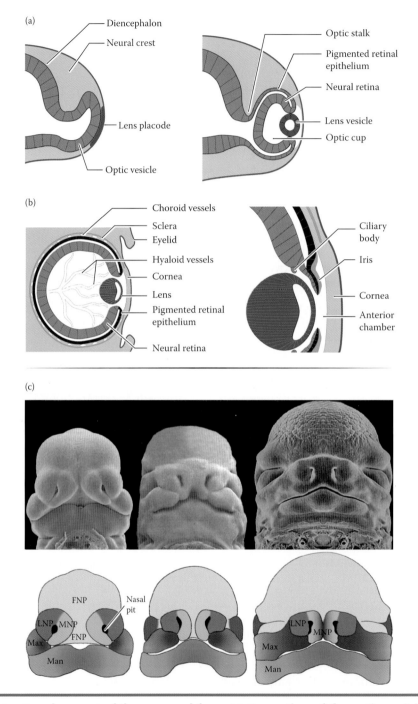

Plate 3.3   Development of the eyes and face. (a) Formation of the optic vesicle, optic cup, and lens vesicle; (b) schematic illustrations showing formation of the cornea and iris; (c) photographs and schemes illustrating the development of the face. MNP: medial nasal prominence, Man: mandibular prominence, Max: maxillary prominence, LNP: lateral nasal prominence, FNP: frontonasal prominence.

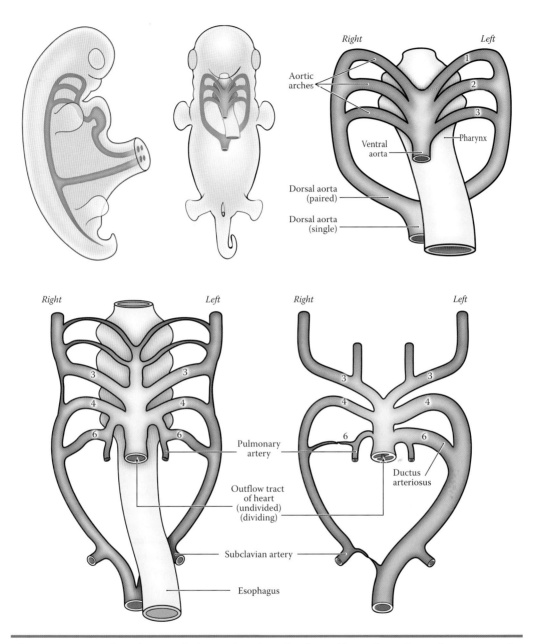

Plate 3.4   Lateral and ventral views showing the formation of aortic arches 1, 2, and 3; each arch passes from ventral to dorsal around the side of the pharynx.

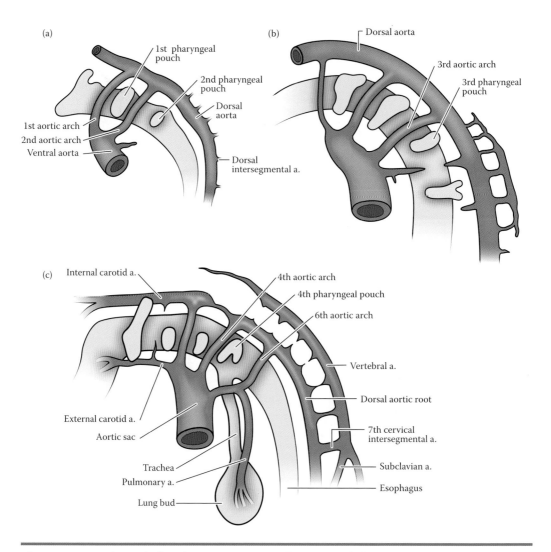

(a)

1st pharyngeal pouch

2nd pharyngeal pouch

Dorsal aorta

1st aortic arch
2nd aortic arch
Ventral aorta

Dorsal intersegmental a.

(b)

Dorsal aorta

3rd aortic arch

3rd pharyngeal pouch

(c)

Internal carotid a.

4th aortic arch

4th pharyngeal pouch

6th aortic arch

Vertebral a.

Dorsal aortic root

External carotid a.

Aortic sac

7th cervical intersegmental a.

Subclavian a.

Trachea

Pulmonary a.

Esophagus

Lung bud

Plate 3.5 Aortic arch development. (a–c) Schematic left lateral views of the progressive formation of aortic arches 1, 2, and 3; note the relation of each aortic arch to the corresponding pharyngeal pouch, which is caudal to the vessel.

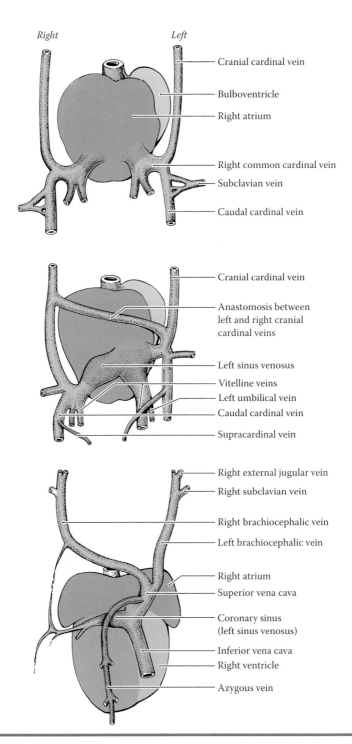

Right   *Left*

Cranial cardinal vein

Bulboventricle

Right atrium

Right common cardinal vein

Subclavian vein

Caudal cardinal vein

Cranial cardinal vein

Anastomosis between
left and right cranial
cardinal veins

Left sinus venosus

Vitelline veins

Left umbilical vein

Caudal cardinal vein

Supracardinal vein

Right external jugular vein

Right subclavian vein

Right brachiocephalic vein

Left brachiocephalic vein

Right atrium

Superior vena cava

Coronary sinus
(left sinus venosus)

Inferior vena cava

Right ventricle

Azygous vein

Plate 3.6   **Dorsal views of the development of the brachiocephalic veins from which subclavians and jugular veins arise. Note the cranial shift in the relative position of the subclavian veins, similar to that occurring during arterial development.**

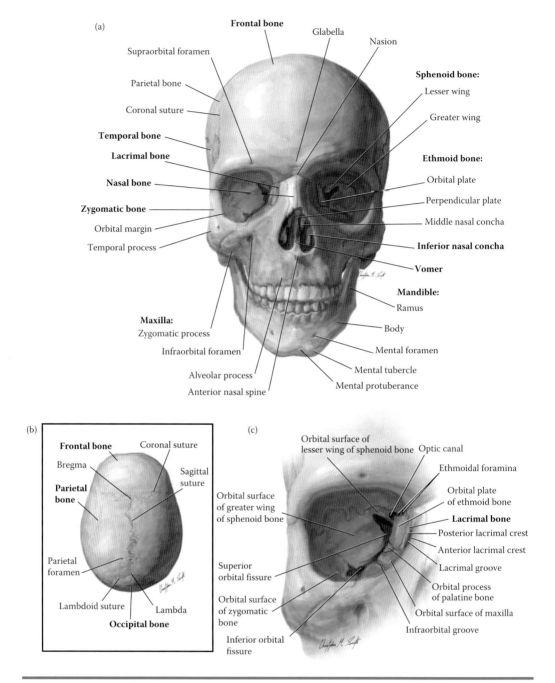

(a)

Frontal bone

Glabella

Nasion

Supraorbital foramen

Parietal bone

Coronal suture

Temporal bone

Lacrimal bone

Nasal bone

Zygomatic bone

Orbital margin

Temporal process

Sphenoid bone:

Lesser wing

Greater wing

Ethmoid bone:

Orbital plate

Perpendicular plate

Middle nasal concha

Inferior nasal concha

Vomer

Mandible:

Ramus

Maxilla:

Zygomatic process

Infraorbital foramen

Alveolar process

Anterior nasal spine

Body

Mental foramen

Mental tubercle

Mental protuberance

(b)

Frontal bone

Coronal suture

Bregma

Sagittal suture

Parietal bone

Parietal foramen

Lambdoid suture

Lambda

Occipital bone

(c)

Orbital surface of lesser wing of sphenoid bone

Optic canal

Ethmoidal foramina

Orbital plate of ethmoid bone

Lacrimal bone

Posterior lacrimal crest

Anterior lacrimal crest

Lacrimal groove

Orbital process of palatine bone

Orbital surface of maxilla

Infraorbital groove

Orbital surface of greater wing of sphenoid bone

Superior orbital fissure

Orbital surface of zygomatic bone

Inferior orbital fissure

Plate 3.7 **The skull. (a) Anterior view; (b) superior view; (c) detail of right orbit, anterior view.**

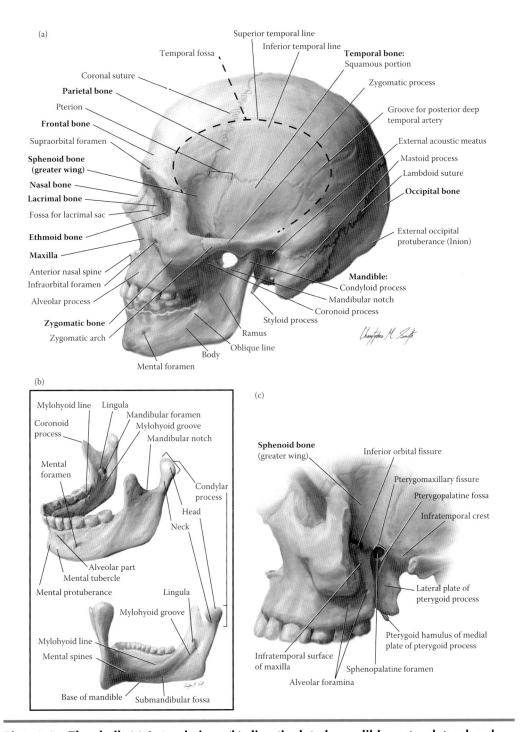

(a)

Superior temporal line
Inferior temporal line
Temporal fossa
Temporal bone:
Squamous portion
Coronal suture
Parietal bone
Zygomatic process
Pterion
Groove for posterior deep
temporal artery
Frontal bone
Supraorbital foramen
External acoustic meatus
Sphenoid bone
(greater wing)
Mastoid process
Nasal bone
Lambdoid suture
Lacrimal bone
Occipital bone
Fossa for lacrimal sac
Ethmoid bone
External occipital
protuberance (Inion)
Maxilla
Anterior nasal spine
Infraorbital foramen
Mandible:
Alveolar process
Condyloid process
Mandibular notch
Zygomatic bone
Coronoid process
Zygomatic arch
Styloid process
Ramus
Oblique line
Body
Mental foramen

(b)

Mylohyoid line   Lingula
Coronoid
process
Mandibular foramen
Mylohyoid groove
Mandibular notch
Mental
foramen
Condylar
process
Head
Neck
Alveolar part
Mental tubercle
Mental protuberance
Lingula
Mylohyoid groove
Mylohyoid line
Mental spines
Base of mandible
Submandibular fossa

(c)

Sphenoid bone
(greater wing)
Inferior orbital fissure
Pterygomaxillary fissure
Pterygopalatine fossa
Infratemporal crest
Lateral plate of
pterygoid process
Pterygoid hamulus of medial
plate of pterygoid process
Infratemporal surface
of maxilla
Sphenopalatine foramen
Alveolar foramina

Plate 3.8   The skull. (a) Lateral view; (b) disarticulated mandible, anterolateral and posterolateral views; (c) infratemporal fossa, lateral view with zygomatic arch and mandible removed.

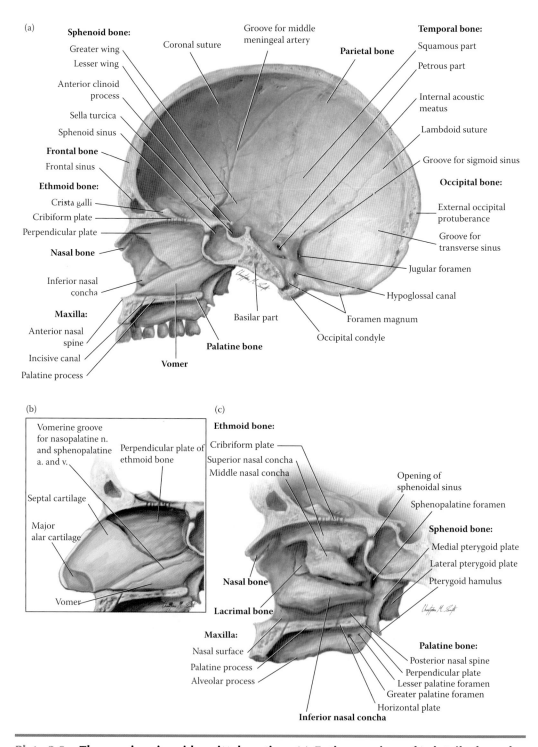

(a)

Sphenoid bone:
Greater wing
Lesser wing
Anterior clinoid process
Sella turcica
Sphenoid sinus

Coronal suture

Groove for middle meningeal artery

Parietal bone

Temporal bone:
Squamous part
Petrous part
Internal acoustic meatus
Lambdoid suture
Groove for sigmoid sinus

Frontal bone
Frontal sinus

Ethmoid bone:
Crista galli
Cribiform plate
Perpendicular plate

Nasal bone

Inferior nasal concha

Maxilla:
Anterior nasal spine
Incisive canal
Palatine process

Vomer

Palatine bone

Basilar part

Occipital condyle

Foramen magnum

Occipital bone:
External occipital protuberance
Groove for transverse sinus
Jugular foramen
Hypoglossal canal

(b)

Vomerine groove for nasopalatine n. and sphenopalatine a. and v.

Perpendicular plate of ethmoid bone

Septal cartilage

Major alar cartilage

Vomer

(c)

Ethmoid bone:
Cribriform plate
Superior nasal concha
Middle nasal concha

Opening of sphenoidal sinus
Sphenopalatine foramen

Sphenoid bone:
Medial pterygoid plate
Lateral pterygoid plate
Pterygoid hamulus

Nasal bone

Lacrimal bone

Maxilla:
Nasal surface
Palatine process
Alveolar process

Palatine bone:
Posterior nasal spine
Perpendicular plate
Lesser palatine foramen
Greater palatine foramen
Horizontal plate

Inferior nasal concha

Plate 3.9 **The cranium in mid-sagittal section. (a) Entire cranium; (b) detail of nasal cavity showing nasal septum; (c) detail of nasal cavity with nasal septum removed to show nasal conchae.**

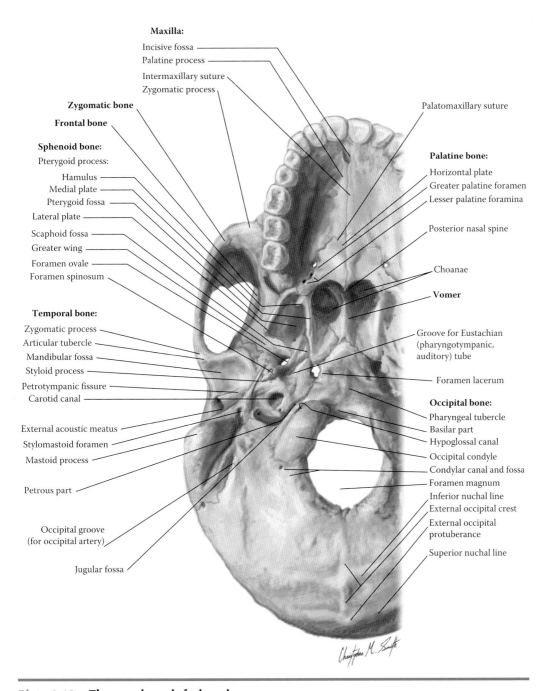

Maxilla:
Incisive fossa
Palatine process
Intermaxillary suture
Zygomatic process

Zygomatic bone

Frontal bone

Sphenoid bone:
Pterygoid process:
Hamulus
Medial plate
Pterygoid fossa
Lateral plate
Scaphoid fossa
Greater wing
Foramen ovale
Foramen spinosum

Temporal bone:
Zygomatic process
Articular tubercle
Mandibular fossa
Styloid process
Petrotympanic fissure
Carotid canal

External acoustic meatus
Stylomastoid foramen
Mastoid process

Petrous part

Occipital groove
(for occipital artery)

Jugular fossa

Palatomaxillary suture

Palatine bone:
Horizontal plate
Greater palatine foramen
Lesser palatine foramina

Posterior nasal spine

Choanae

Vomer

Groove for Eustachian
(pharyngotympanic,
auditory) tube

Foramen lacerum

Occipital bone:
Pharyngeal tubercle
Basilar part
Hypoglossal canal
Occipital condyle
Condylar canal and fossa
Foramen magnum
Inferior nuchal line
External occipital crest
External occipital
protuberance
Superior nuchal line

Plate 3.10   The cranium, inferior view.

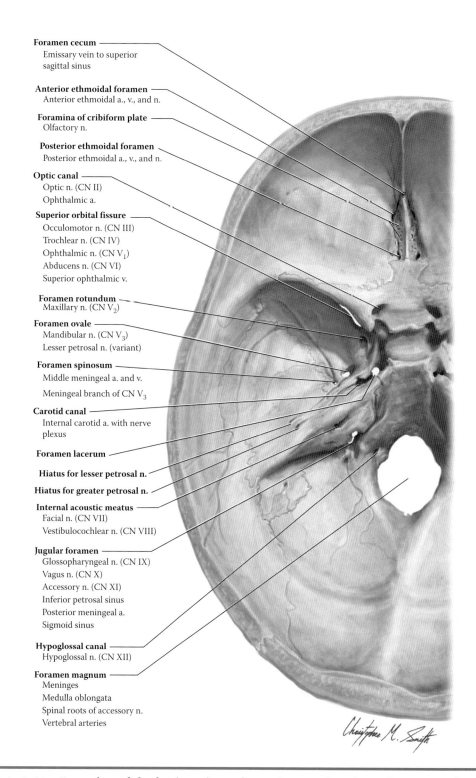

Foramen cecum
  Emissary vein to superior
  sagittal sinus

Anterior ethmoidal foramen
  Anterior ethmoidal a., v., and n.

Foramina of cribiform plate
  Olfactory n.

Posterior ethmoidal foramen
  Posterior ethmoidal a., v., and n.

Optic canal
  Optic n. (CN II)
  Ophthalmic a.

Superior orbital fissure
  Occulomotor n. (CN III)
  Trochlear n. (CN IV)
  Ophthalmic n. (CN V_1)
  Abducens n. (CN VI)
  Superior ophthalmic v.

Foramen rotundum
  Maxillary n. (CN V_2)

Foramen ovale
  Mandibular n. (CN V_3)
  Lesser petrosal n. (variant)

Foramen spinosum
  Middle meningeal a. and v.
  Meningeal branch of CN V_3

Carotid canal
  Internal carotid a. with nerve
  plexus

Foramen lacerum

Hiatus for lesser petrosal n.

Hiatus for greater petrosal n.

Internal acoustic meatus
  Facial n. (CN VII)
  Vestibulocochlear n. (CN VIII)

Jugular foramen
  Glossopharyngeal n. (CN IX)
  Vagus n. (CN X)
  Accessory n. (CN XI)
  Inferior petrosal sinus
  Posterior meningeal a.
  Sigmoid sinus

Hypoglossal canal
  Hypoglossal n. (CN XII)

Foramen magnum
  Meninges
  Medulla oblongata
  Spinal roots of accessory n.
  Vertebral arteries

Plate 3.11   Foramina of the basicranium, shown in superior view of transversely-sectioned cranium.

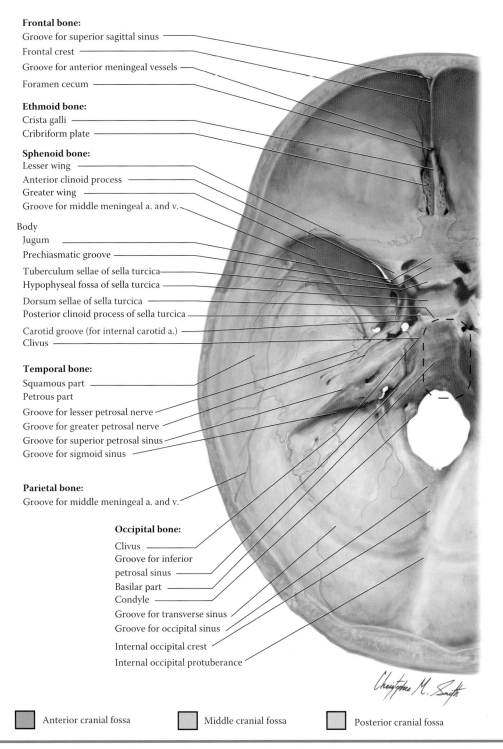

Frontal bone:
Groove for superior sagittal sinus
Frontal crest
Groove for anterior meningeal vessels
Foramen cecum

Ethmoid bone:
Crista galli
Cribriform plate

Sphenoid bone:
Lesser wing
Anterior clinoid process
Greater wing
Groove for middle meningeal a. and v.

Body
Jugum
Prechiasmatic groove

Tuberculum sellae of sella turcica
Hypophyseal fossa of sella turcica

Dorsum sellae of sella turcica
Posterior clinoid process of sella turcica

Carotid groove (for internal carotid a.)
Clivus

Temporal bone:
Squamous part
Petrous part
Groove for lesser petrosal nerve
Groove for greater petrosal nerve
Groove for superior petrosal sinus
Groove for sigmoid sinus

Parietal bone:
Groove for middle meningeal a. and v.

Occipital bone:
Clivus
Groove for inferior
petrosal sinus
Basilar part
Condyle
Groove for transverse sinus
Groove for occipital sinus
Internal occipital crest
Internal occipital protuberance

Anterior cranial fossa   Middle cranial fossa   Posterior cranial fossa

Plate 3.12   The cranial fossae, shown in superior view of transversely-sectioned cranium.

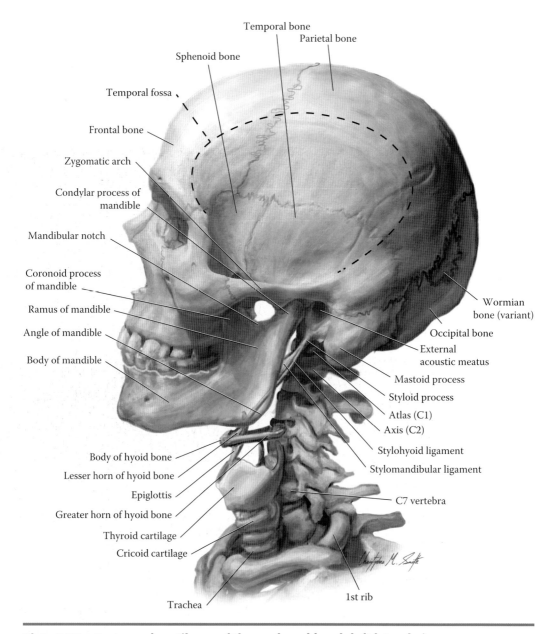

Temporal bone
Parietal bone
Sphenoid bone
Temporal fossa
Frontal bone
Zygomatic arch
Condylar process of mandible
Mandibular notch
Coronoid process of mandible
Ramus of mandible
Angle of mandible
Body of mandible
Body of hyoid bone
Lesser horn of hyoid bone
Epiglottis
Greater horn of hyoid bone
Thyroid cartilage
Cricoid cartilage
Trachea
Wormian bone (variant)
Occipital bone
External acoustic meatus
Mastoid process
Styloid process
Atlas (C1)
Axis (C2)
Stylohyoid ligament
Stylomandibular ligament
C7 vertebra
1st rib

Plate 3.13   Bones and cartilages of the neck and head, left lateral view.

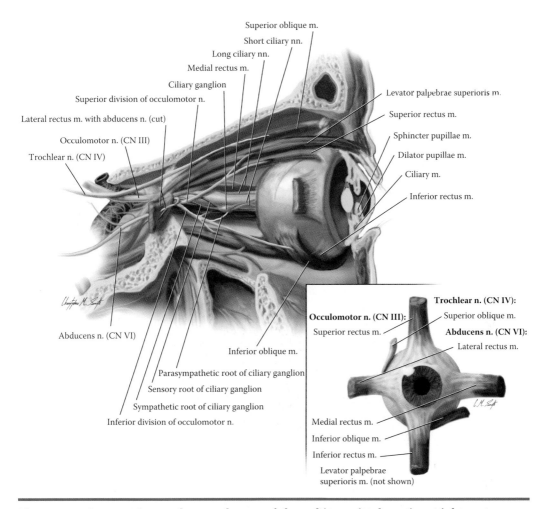

Superior oblique m.
Short ciliary nn.
Long ciliary nn.
Medial rectus m.
Ciliary ganglion
Superior division of occulomotor n.
Lateral rectus m. with abducens n. (cut)
Occulomotor n. (CN III)
Trochlear n. (CN IV)

Levator palpebrae superioris m.
Superior rectus m.
Sphincter pupillae m.
Dilator pupillae m.
Ciliary m.
Inferior rectus m.

Abducens n. (CN VI)
Inferior oblique m.
Parasympathetic root of ciliary ganglion
Sensory root of ciliary ganglion
Sympathetic root of ciliary ganglion
Inferior division of occulomotor n.

Trochlear n. (CN IV):
Superior oblique m.
Occulomotor n. (CN III):
Superior rectus m.
Abducens n. (CN VI):
Lateral rectus m.

Medial rectus m.
Inferior oblique m.
Inferior rectus m.
Levator palpebrae superioris m. (not shown)

Plate 3.14 **Innervation and musculature of the orbit, sagittal section (right eye). Blue and green colors correspond to branches of the trigeminal nerve (CN V). Inset: Anterior view of right eye showing external ocular muscles with rectus muscles reflected.**

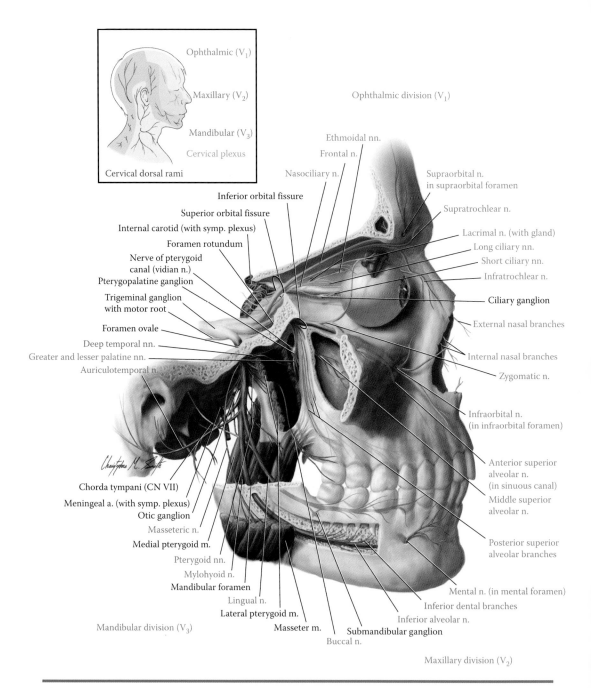

Ophthalmic (V₁)

Maxillary (V₂)

Mandibular (V₃)

Cervical plexus

Cervical dorsal rami

Ophthalmic division (V₁)

Ethmoidal nn.

Frontal n.

Nasociliary n.

Inferior orbital fissure

Superior orbital fissure

Internal carotid (with symp. plexus)

Foramen rotundum

Nerve of pterygoid canal (vidian n.)

Pterygopalatine ganglion

Trigeminal ganglion with motor root

Foramen ovale

Deep temporal nn.

Greater and lesser palatine nn.

Auriculotemporal n.

Supraorbital n. in supraorbital foramen

Supratrochlear n.

Lacrimal n. (with gland)

Long ciliary nn.

Short ciliary nn.

Infratrochlear n.

Ciliary ganglion

External nasal branches

Internal nasal branches

Zygomatic n.

Infraorbital n. (in infraorbital foramen)

Anterior superior alveolar n. (in sinuous canal)

Middle superior alveolar n.

Posterior superior alveolar branches

Chorda tympani (CN VII)

Meningeal a. (with symp. plexus)

Otic ganglion

Masseteric n.

Medial pterygoid m.

Pterygoid nn.

Mylohyoid n.

Mandibular foramen

Lingual n.

Lateral pterygoid m.

Mandibular division (V₃)

Masseter m.

Buccal n.

Submandibular ganglion

Mental n. (in mental foramen)

Inferior dental branches

Inferior alveolar n.

Maxillary division (V₂)

Plate 3.15 Divisions of the trigeminal nerve (CN V), left lateral view with part of the cranium and mandible removed. Inset: Cutaneous innervation of the head.

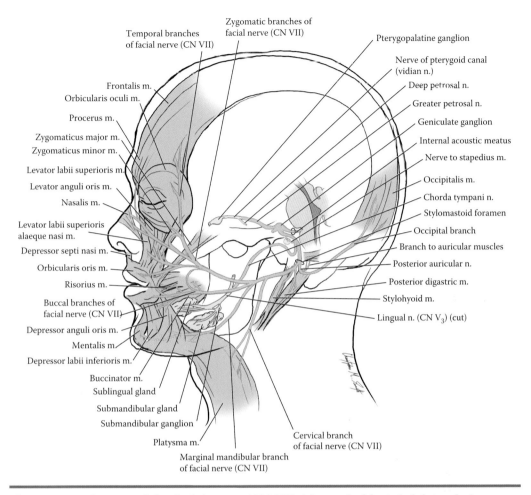

Zygomatic branches of
facial nerve (CN VII)

Temporal branches
of facial nerve (CN VII)

Pterygopalatine ganglion

Nerve of pterygoid canal
(vidian n.)

Frontalis m.

Deep petrosal n.

Orbicularis oculi m.

Greater petrosal n.

Procerus m.

Geniculate ganglion

Zygomaticus major m.

Internal acoustic meatus

Zygomaticus minor m.

Nerve to stapedius m.

Levator labii superioris m.

Occipitalis m.

Levator anguli oris m.

Chorda tympani n.

Nasalis m.

Stylomastoid foramen

Levator labii superioris
alaeque nasi m.

Occipital branch

Branch to auricular muscles

Depressor septi nasi m.

Posterior auricular n.

Orbicularis oris m.

Posterior digastric m.

Risorius m.

Stylohyoid m.

Buccal branches of
facial nerve (CN VII)

Lingual n. (CN V$_3$) (cut)

Depressor anguli oris m.

Mentalis m.

Depressor labii inferioris m.

Buccinator m.

Sublingual gland

Submandibular gland

Submandibular ganglion

Cervical branch
of facial nerve (CN VII)

Platysma m.

Marginal mandibular branch
of facial nerve (CN VII)

Plate 3.16   **Diagram of the facial nerve (CN VII) (shown in blue), left lateral view.**

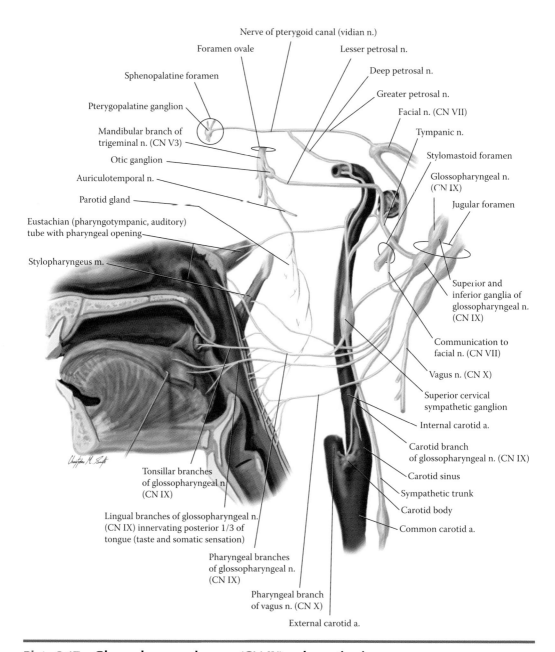

Nerve of pterygoid canal (vidian n.)

Foramen ovale

Lesser petrosal n.

Sphenopalatine foramen

Deep petrosal n.

Greater petrosal n.

Pterygopalatine ganglion

Facial n. (CN VII)

Mandibular branch of
trigeminal n. (CN V3)

Tympanic n.

Otic ganglion

Stylomastoid foramen

Auriculotemporal n.

Glossopharyngeal n.
(CN IX)

Parotid gland

Jugular foramen

Eustachian (pharyngotympanic, auditory)
tube with pharyngeal opening

Stylopharyngeus m.

Superior and
inferior ganglia of
glossopharyngeal n.
(CN IX)

Communication to
facial n. (CN VII)

Vagus n. (CN X)

Superior cervical
sympathetic ganglion

Internal carotid a.

Carotid branch
of glossopharyngeal n. (CN IX)

Carotid sinus

Sympathetic trunk

Carotid body

Common carotid a.

Tonsillar branches
of glossopharyngeal n.
(CN IX)

Lingual branches of glossopharyngeal n.
(CN IX) innervating posterior 1/3 of
tongue (taste and somatic sensation)

Pharyngeal branches
of glossopharyngeal n.
(CN IX)

Pharyngeal branch
of vagus n. (CN X)

External carotid a.

Plate 3.17   Glossopharyngeal nerve (CN IX), schematic view.

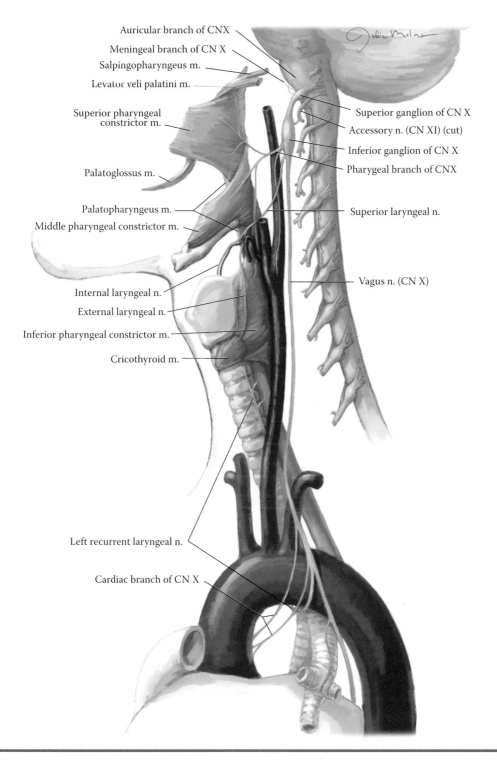

Auricular branch of CNX
Meningeal branch of CN X
Salpingopharyngeus m.
Levator veli palatini m.
Superior pharyngeal constrictor m.
Palatoglossus m.
Palatopharyngeus m.
Middle pharyngeal constrictor m.
Internal laryngeal n.
External laryngeal n.
Inferior pharyngeal constrictor m.
Cricothyroid m.
Left recurrent laryngeal n.
Cardiac branch of CN X

Superior ganglion of CN X
Accessory n. (CN XI) (cut)
Inferior ganglion of CN X
Pharygeal branch of CNX
Superior laryngeal n.
Vagus n. (CN X)

Plate 3.18   Vagus nerve (CN X) in the head and neck, left lateral view.

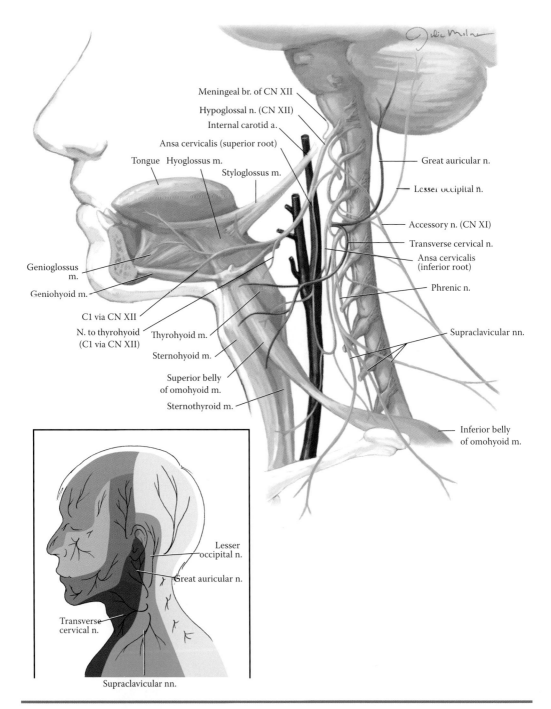

Meningeal br. of CN XII
Hypoglossal n. (CN XII)
Internal carotid a.
Ansa cervicalis (superior root)
Tongue Hyoglossus m.
Styloglossus m.
Great auricular n.
Lesser occipital n.
Accessory n. (CN XI)
Transverse cervical n.
Ansa cervicalis (inferior root)
Genioglossus m.
Phrenic n.
Geniohyoid m.
C1 via CN XII
N. to thyrohyoid (C1 via CN XII)
Thyrohyoid m.
Sternohyoid m.
Supraclavicular nn.
Superior belly of omohyoid m.
Sternothyroid m.
Inferior belly of omohyoid m.

Lesser occipital n.
Great auricular n.
Transverse cervical n.
Supraclavicular nn.

Plate 3.19   The cervical plexus and hypoglossal nerve (CN XII), left lateral view. Inset: Cutaneous innervation of the head and neck by branches of the cervical plexus.

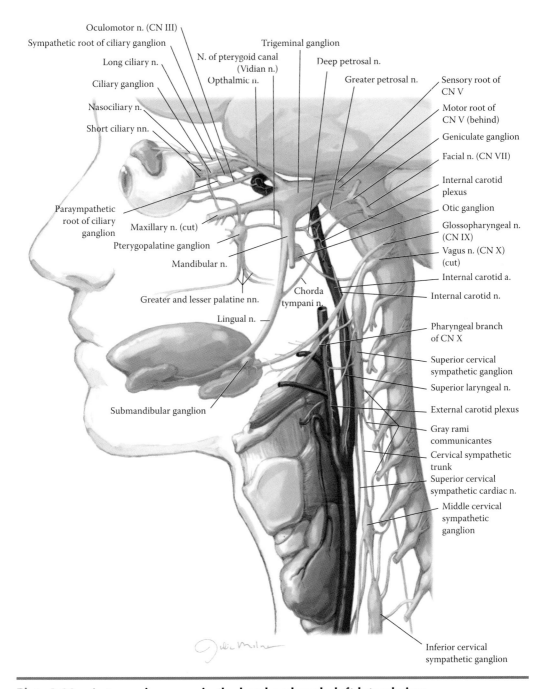

Oculomotor n. (CN III)

Sympathetic root of ciliary ganglion

Long ciliary n.

Ciliary ganglion

Nasociliary n.

Short ciliary nn.

N. of pterygoid canal
(Vidian n.)

Opthalmic n.

Trigeminal ganglion

Deep petrosal n.

Greater petrosal n.

Sensory root of
CN V

Motor root of
CN V (behind)

Geniculate ganglion

Facial n. (CN VII)

Internal carotid
plexus

Paraympathetic
root of ciliary
ganglion

Maxillary n. (cut)

Pterygopalatine ganglion

Mandibular n.

Otic ganglion

Glossopharyngeal n.
(CN IX)

Vagus n. (CN X)
(cut)

Internal carotid a.

Internal carotid n.

Greater and lesser palatine nn.

Chorda
tympani n.

Lingual n.

Pharyngeal branch
of CN X

Superior cervical
sympathetic ganglion

Superior laryngeal n.

External carotid plexus

Gray rami
communicantes

Cervical sympathetic
trunk

Superior cervical
sympathetic cardiac n.

Submandibular ganglion

Middle cervical
sympathetic
ganglion

Inferior cervical
sympathetic ganglion

Plate 3.20   **Autonomic nerves in the head and neck, left lateral view.**

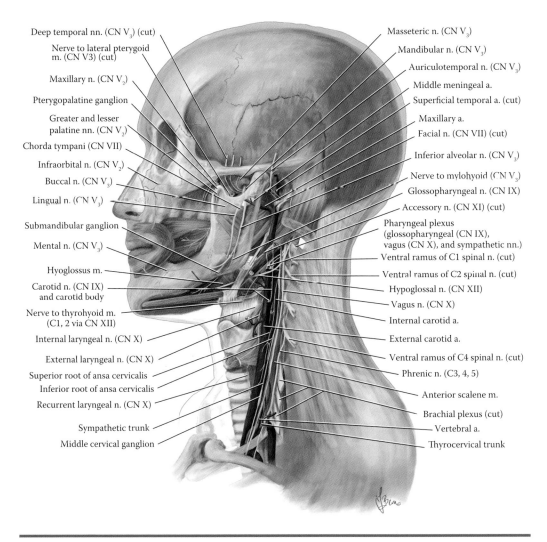

Deep temporal nn. (CN V₃) (cut)

Nerve to lateral pterygoid m. (CN V3) (cut)

Maxillary n. (CN V₂)

Pterygopalatine ganglion

Greater and lesser palatine nn. (CN V₂)

Chorda tympani (CN VII)

Infraorbital n. (CN V₂)

Buccal n. (CN V₃)

Lingual n. (CN V₃)

Submandibular ganglion

Mental n. (CN V₃)

Hyoglossus m.

Carotid n. (CN IX) and carotid body

Nerve to thyrohyoid m. (C1, 2 via CN XII)

Internal laryngeal n. (CN X)

External laryngeal n. (CN X)

Superior root of ansa cervicalis

Inferior root of ansa cervicalis

Recurrent laryngeal n. (CN X)

Sympathetic trunk

Middle cervical ganglion

Masseteric n. (CN V₃)

Mandibular n. (CN V₃)

Auriculotemporal n. (CN V₃)

Middle meningeal a.

Superficial temporal a. (cut)

Maxillary a.

Facial n. (CN VII) (cut)

Inferior alveolar n. (CN V₃)

Nerve to mylohyoid (CN V₃)

Glossopharyngeal n. (CN IX)

Accessory n. (CN XI) (cut)

Pharyngeal plexus (glossopharyngeal (CN IX), vagus (CN X), and sympathetic nn.)

Ventral ramus of C1 spinal n. (cut)

Ventral ramus of C2 spinal n. (cut)

Hypoglossal n. (CN XII)

Vagus n. (CN X)

Internal carotid a.

External carotid a.

Ventral ramus of C4 spinal n. (cut)

Phrenic n. (C3, 4, 5)

Anterior scalene m.

Brachial plexus (cut)

Vertebral a.

Thyrocervical trunk

Plate 3.21 Superficial nerves of the head and neck, left lateral view. Neck muscles and part of the mandible are rendered transparent.

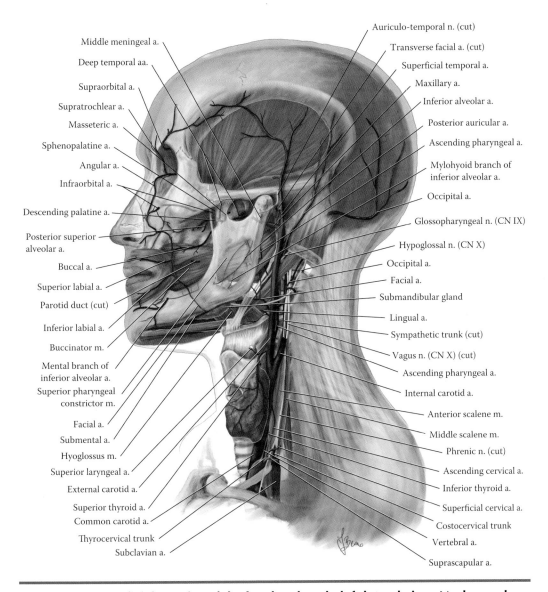

Middle meningeal a.

Deep temporal aa.

Supraorbital a.

Supratrochlear a.

Masseteric a.

Sphenopalatine a.

Angular a.

Infraorbital a.

Descending palatine a.

Posterior superior
alveolar a.

Buccal a.

Superior labial a.

Parotid duct (cut)

Inferior labial a.

Buccinator m.

Mental branch of
inferior alveolar a.

Superior pharyngeal
constrictor m.

Facial a.

Submental a.

Hyoglossus m.

Superior laryngeal a.

External carotid a.

Superior thyroid a.

Common carotid a.

Thyrocervical trunk

Subclavian a.

Auriculo-temporal n. (cut)

Transverse facial a. (cut)

Superficial temporal a.

Maxillary a.

Inferior alveolar a.

Posterior auricular a.

Ascending pharyngeal a.

Mylohyoid branch of
inferior alveolar a.

Occipital a.

Glossopharyngeal n. (CN IX)

Hypoglossal n. (CN X)

Occipital a.

Facial a.

Submandibular gland

Lingual a.

Sympathetic trunk (cut)

Vagus n. (CN X) (cut)

Ascending pharyngeal a.

Internal carotid a.

Anterior scalene m.

Middle scalene m.

Phrenic n. (cut)

Ascending cervical a.

Inferior thyroid a.

Superficial cervical a.

Costocervical trunk

Vertebral a.

Suprascapular a.

**Plate 3.22 Superficial arteries of the head and neck, left lateral view. Neck muscles
and part of the mandible are rendered transparent.**

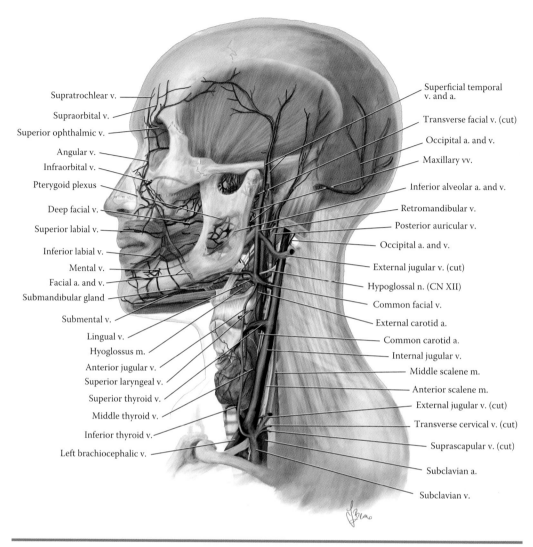

Supratrochlear v.

Supraorbital v.

Superior ophthalmic v.

Angular v.

Infraorbital v.

Pterygoid plexus

Deep facial v.

Superior labial v.

Inferior labial v.

Mental v.

Facial a. and v.

Submandibular gland

Submental v.

Lingual v.

Hyoglossus m.

Anterior jugular v.

Superior laryngeal v.

Superior thyroid v.

Middle thyroid v.

Inferior thyroid v.

Left brachiocephalic v.

Superficial temporal v. and a.

Transverse facial v. (cut)

Occipital a. and v.

Maxillary vv.

Inferior alveolar a. and v.

Retromandibular v.

Posterior auricular v.

Occipital a. and v.

External jugular v. (cut)

Hypoglossal n. (CN XII)

Common facial v.

External carotid a.

Common carotid a.

Internal jugular v.

Middle scalene m.

Anterior scalene m.

External jugular v. (cut)

Transverse cervical v. (cut)

Suprascapular v. (cut)

Subclavian a.

Subclavian v.

Plate 3.23   Superficial veins of the head and neck, left lateral view. Neck muscles and part of the mandible are rendered transparent.

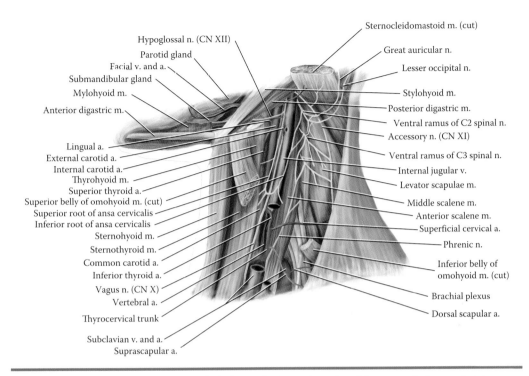

Sternocleidomastoid m. (cut)

Hypoglossal n. (CN XII)

Parotid gland

Great auricular n.

Facial v. and a.

Lesser occipital n.

Submandibular gland

Mylohyoid m.

Stylohyoid m.

Anterior digastric m.

Posterior digastric m.

Ventral ramus of C2 spinal n.

Accessory n. (CN XI)

Lingual a.

External carotid a.

Ventral ramus of C3 spinal n.

Internal carotid a.

Internal jugular v.

Thyrohyoid m.

Levator scapulae m.

Superior thyroid a.

Superior belly of omohyoid m. (cut)

Middle scalene m.

Superior root of ansa cervicalis

Anterior scalene m.

Inferior root of ansa cervicalis

Superficial cervical a.

Sternohyoid m.

Phrenic n.

Sternothyroid m.

Common carotid a.

Inferior belly of omohyoid m. (cut)

Inferior thyroid a.

Vagus n. (CN X)

Brachial plexus

Vertebral a.

Dorsal scapular a.

Thyrocervical trunk

Subclavian v. and a.

Suprascapular a.

Plate 3.24   The cervical region, left lateral view. Omohyoid cut and sternocleidomastoid cut and reflected.

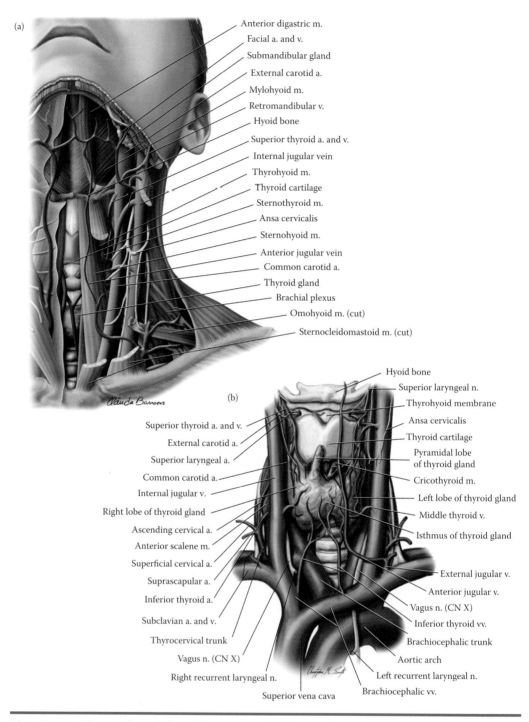

(a)

Anterior digastric m.
Facial a. and v.
Submandibular gland
External carotid a.
Mylohyoid m.
Retromandibular v.
Hyoid bone
Superior thyroid a. and v.
Internal jugular vein
Thyrohyoid m.
Thyroid cartilage
Sternothyroid m.
Ansa cervicalis
Sternohyoid m.
Anterior jugular vein
Common carotid a.
Thyroid gland
Brachial plexus
Omohyoid m. (cut)
Sternocleidomastoid m. (cut)

(b)

Hyoid bone
Superior laryngeal n.
Thyrohyoid membrane
Ansa cervicalis
Thyroid cartilage
Pyramidal lobe of thyroid gland
Cricothyroid m.
Left lobe of thyroid gland
Middle thyroid v.
Isthmus of thyroid gland
External jugular v.
Anterior jugular v.
Vagus n. (CN X)
Inferior thyroid vv.
Brachiocephalic trunk
Aortic arch
Left recurrent laryngeal n.
Brachiocephalic vv.

Superior thyroid a. and v.
External carotid a.
Superior laryngeal a.
Common carotid a.
Internal jugular v.
Right lobe of thyroid gland
Ascending cervical a.
Anterior scalene m.
Superficial cervical a.
Suprascapular a.
Inferior thyroid a.
Subclavian a. and v.
Thyrocervical trunk
Vagus n. (CN X)
Right recurrent laryngeal n.
Superior vena cava

Plate 3.25   **The neck and thyroid. (a) Superficial vessels and nerves of neck, anterior view; (b) vasculature of the thyroid gland, anterior view.**

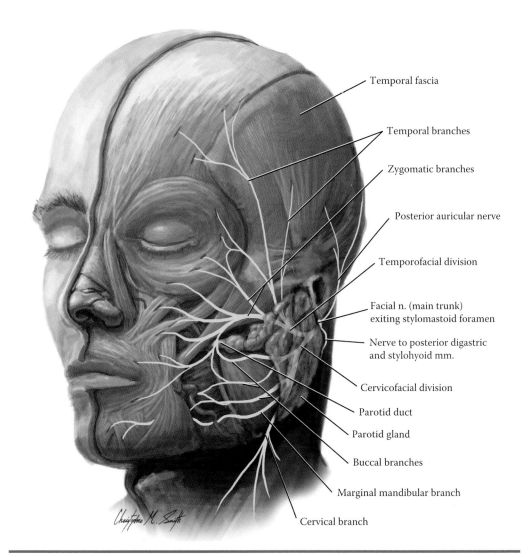

Temporal fascia

Temporal branches

Zygomatic branches

Posterior auricular nerve

Temporofacial division

Facial n. (main trunk)
exiting stylomastoid foramen

Nerve to posterior digastric
and stylohyoid mm.

Cervicofacial division

Parotid duct

Parotid gland

Buccal branches

Marginal mandibular branch

Cervical branch

Plate 3.26   Branches of the facial nerve (CN VII), anterolateral view.

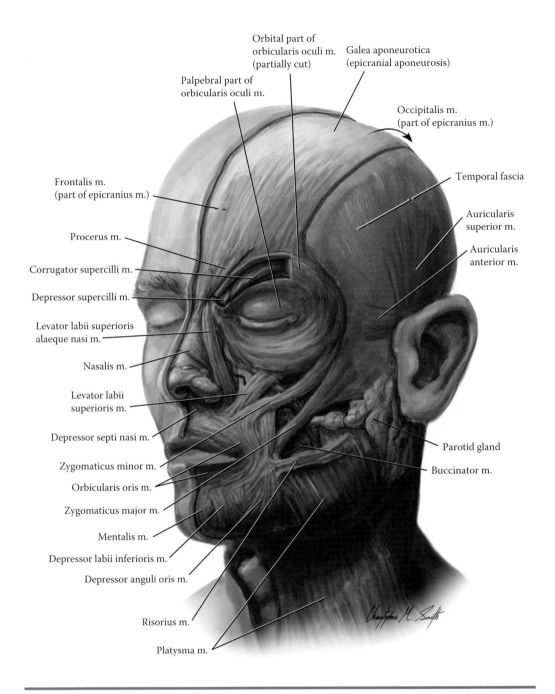

Orbital part of
orbicularis oculi m.
(partially cut)

Palpebral part of
orbicularis oculi m.

Galea aponeurotica
(epicranial aponeurosis)

Occipitalis m.
(part of epicranius m.)

Frontalis m.
(part of epicranius m.)

Temporal fascia

Auricularis
superior m.

Procerus m.

Auricularis
anterior m.

Corrugator supercilli m.

Depressor supercilli m.

Levator labii superioris
alaeque nasi m.

Nasalis m.

Levator labii
superioris m.

Depressor septi nasi m.

Parotid gland

Zygomaticus minor m.

Buccinator m.

Orbicularis oris m.

Zygomaticus major m.

Mentalis m.

Depressor labii inferioris m.

Depressor anguli oris m.

Risorius m.

Platysma m.

Plate 3.27   **Muscles of facial expression, anterolateral view. All muscles shown are innervated by the facial nerve (CN VII).**

(a)

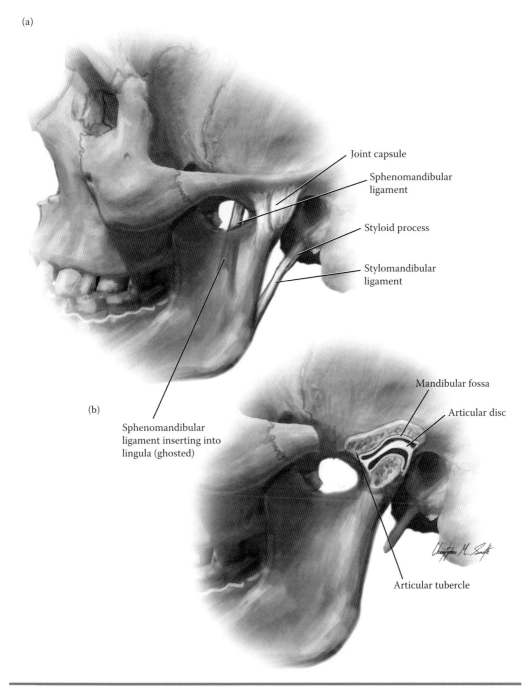

Joint capsule

Sphenomandibular ligament

Styloid process

Stylomandibular ligament

Mandibular fossa

Articular disc

(b)

Sphenomandibular ligament inserting into lingula (ghosted)

Articular tubercle

Plate 3.28   The temporomandibular joint. (a) Left lateral view; (b) left lateral view with part of the zygomatic arch and mandible removed to show the joint cavity.

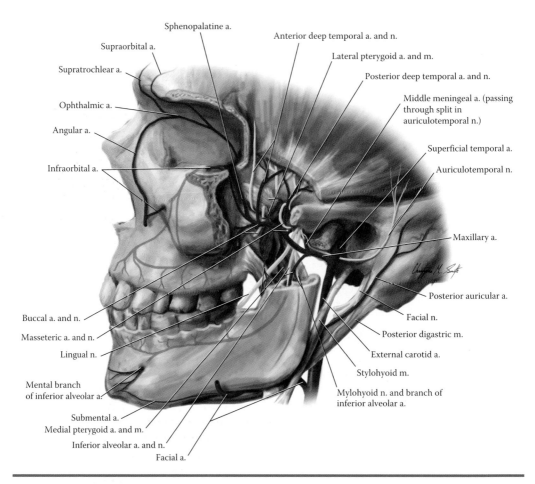

Plate 3.29   **The maxillary artery, left lateral view with zygomatic arch and part of mandible removed to show branches of the maxillary artery and nerves in the infratemporal fossa.**

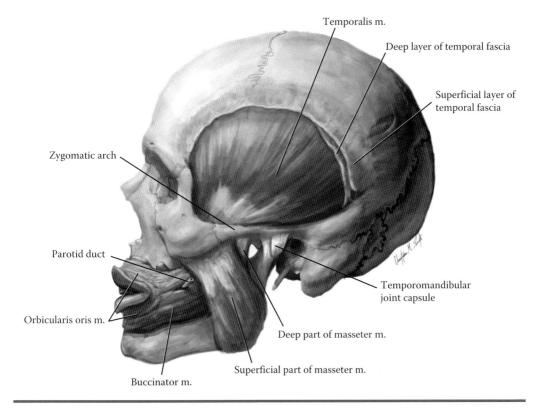

Temporalis m.

Deep layer of temporal fascia

Superficial layer of temporal fascia

Zygomatic arch

Parotid duct

Orbicularis oris m.

Temporomandibular joint capsule

Deep part of masseter m.

Superficial part of masseter m.

Buccinator m.

Plate 3.30 Muscles of mastication, left lateral view with temporal fascia removed to show temporalis muscle.

(a)

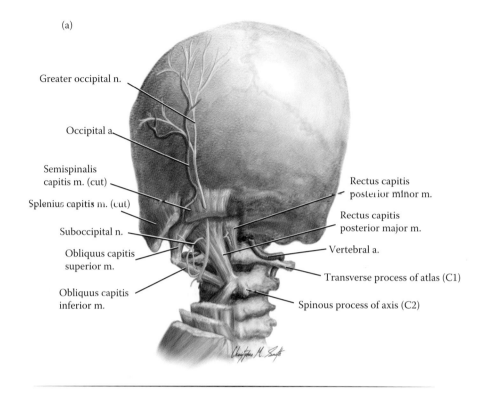

Greater occipital n.

Occipital a.

Semispinalis
capitis m. (cut)

Splenius capitis m. (cut)

Suboccipital n.

Obliquus capitis
superior m.

Obliquus capitis
inferior m.

Rectus capitis
posterior minor m.

Rectus capitis
posterior major m.

Vertebral a.

Transverse process of atlas (C1)

Spinous process of axis (C2)

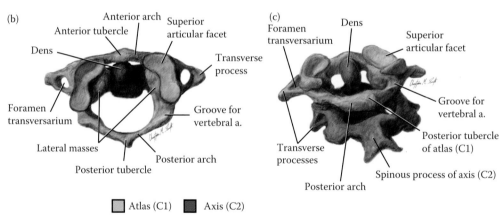

(b)

Anterior arch   Superior
Anterior tubercle   articular facet
Dens

Foramen
transversarium

Lateral masses

Posterior tubercle

Transverse
process

Groove for
vertebral a.

Posterior arch

(c)
Foramen
transversarium

Dens

Superior
articular facet

Groove for
vertebral a.

Posterior tubercle
of atlas (C1)

Transverse
processes

Spinous process of axis (C2)

Posterior arch

☐ Atlas (C1)   ◼ Axis (C2)

Plate 3.31   **The suboccipital region. (a) Posterior view; (b) atlas/axis (C1/C2)
articulation, superior view; (c) atlas/axis (C1/C2) articulation, posterolateral view.**

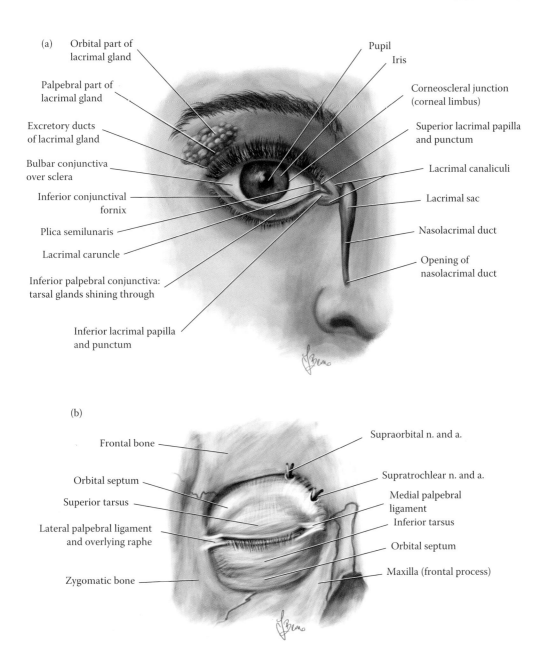

(a) Orbital part of lacrimal gland

Palpebral part of lacrimal gland

Excretory ducts of lacrimal gland

Bulbar conjunctiva over sclera

Inferior conjunctival fornix

Plica semilunaris

Lacrimal caruncle

Inferior palpebral conjunctiva: tarsal glands shining through

Inferior lacrimal papilla and punctum

Pupil

Iris

Corneoscleral junction (corneal limbus)

Superior lacrimal papilla and punctum

Lacrimal canaliculi

Lacrimal sac

Nasolacrimal duct

Opening of nasolacrimal duct

(b)

Frontal bone

Orbital septum

Superior tarsus

Lateral palpebral ligament and overlying raphe

Zygomatic bone

Supraorbital n. and a.

Supratrochlear n. and a.

Medial palpebral ligament

Inferior tarsus

Orbital septum

Maxilla (frontal process)

Plate 3.32   The orbital region, anterior view (right side). (a) External structures of the eye with lower lid retracted to show tarsal glands; (b) skin removed to show orbital septum and palpebral ligaments.

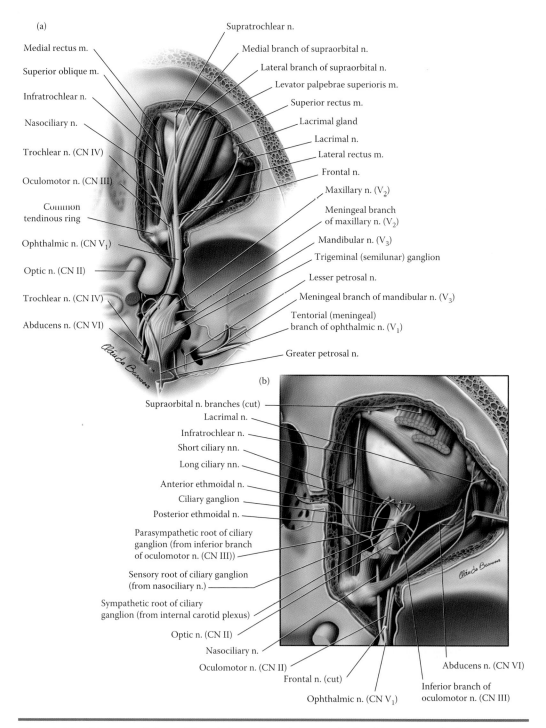

(a)

Supratrochlear n.

Medial rectus m.

Superior oblique m.

Infratrochlear n.

Nasociliary n.

Trochlear n. (CN IV)

Oculomotor n. (CN III)

Common tendinous ring

Ophthalmic n. (CN V₁)

Optic n. (CN II)

Trochlear n. (CN IV)

Abducens n. (CN VI)

Medial branch of supraorbital n.

Lateral branch of supraorbital n.

Levator palpebrae superioris m.

Superior rectus m.

Lacrimal gland

Lacrimal n.

Lateral rectus m.

Frontal n.

Maxillary n. (V₂)

Meningeal branch of maxillary n. (V₂)

Mandibular n. (V₃)

Trigeminal (semilunar) ganglion

Lesser petrosal n.

Meningeal branch of mandibular n. (V₃)

Tentorial (meningeal) branch of ophthalmic n. (V₁)

Greater petrosal n.

(b)

Supraorbital n. branches (cut)

Lacrimal n.

Infratrochlear n.

Short ciliary nn.

Long ciliary nn.

Anterior ethmoidal n.

Ciliary ganglion

Posterior ethmoidal n.

Parasympathetic root of ciliary ganglion (from inferior branch of oculomotor n. (CN III))

Sensory root of ciliary ganglion (from nasociliary n.)

Sympathetic root of ciliary ganglion (from internal carotid plexus)

Optic n. (CN II)

Nasociliary n.

Oculomotor n. (CN II)

Frontal n. (cut)

Ophthalmic n. (CN V₁)

Abducens n. (CN VI)

Inferior branch of oculomotor n. (CN III)

Plate 3.33   Nerves of the orbit. (a) Superior view, roof of orbit removed; (b) superior view, deep dissection.

(a)

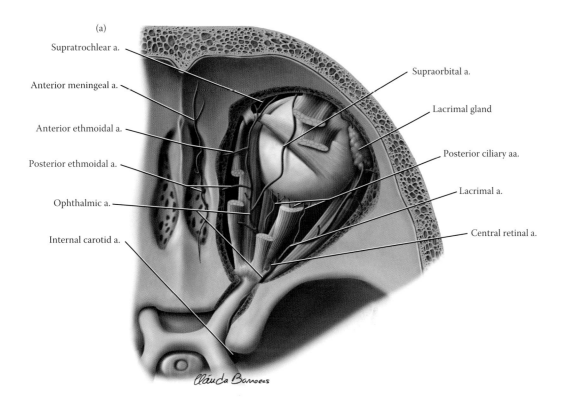

Supratrochlear a.

Anterior meningeal a.

Anterior ethmoidal a.

Posterior ethmoidal a.

Ophthalmic a.

Internal carotid a.

Supraorbital a.

Lacrimal gland

Posterior ciliary aa.

Lacrimal a.

Central retinal a.

(b)

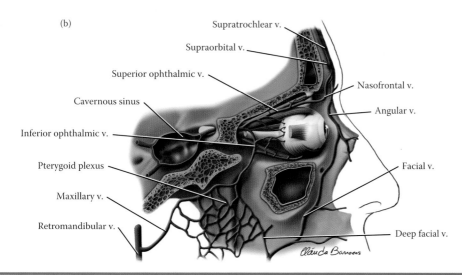

Supratrochlear v.

Supraorbital v.

Superior ophthalmic v.

Cavernous sinus

Inferior ophthalmic v.

Pterygoid plexus

Maxillary v.

Retromandibular v.

Nasofrontal v.

Angular v.

Facial v.

Deep facial v.

Plate 3.34 Arteries and veins of orbit. (a) Superior view, roof of orbit removed; (b) right lateral view.

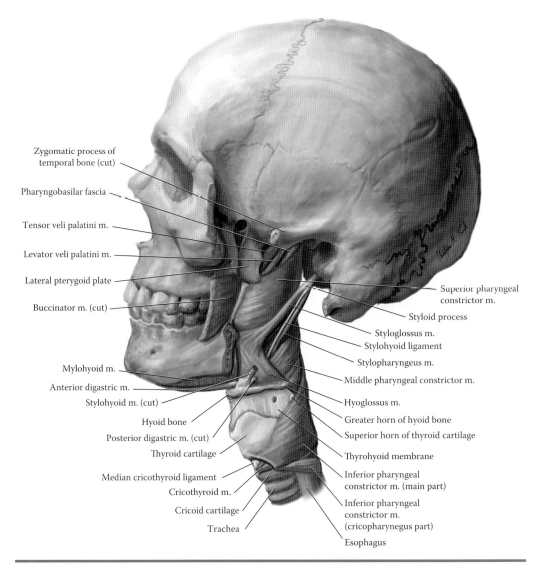

Zygomatic process of
temporal bone (cut)

Pharyngobasilar fascia

Tensor veli palatini m.

Levator veli palatini m.

Lateral pterygoid plate

Buccinator m. (cut)

Mylohyoid m.

Anterior digastric m.

Stylohyoid m. (cut)

Hyoid bone

Posterior digastric m. (cut)

Thyroid cartilage

Median cricothyroid ligament

Cricothyroid m.

Cricoid cartilage

Trachea

Superior pharyngeal
constrictor m.

Styloid process

Styloglossus m.

Stylohyoid ligament

Stylopharyngeus m.

Middle pharyngeal constrictor m.

Hyoglossus m.

Greater horn of hyoid bone

Superior horn of thyroid cartilage

Thyrohyoid membrane

Inferior pharyngeal
constrictor m. (main part)

Inferior pharyngeal
constrictor m.
(cricopharynegus part)

Esophagus

**Plate 3.35   Musculature of the pharynx and larynx, left lateral view with zygomatic
arch and part of mandible removed.**

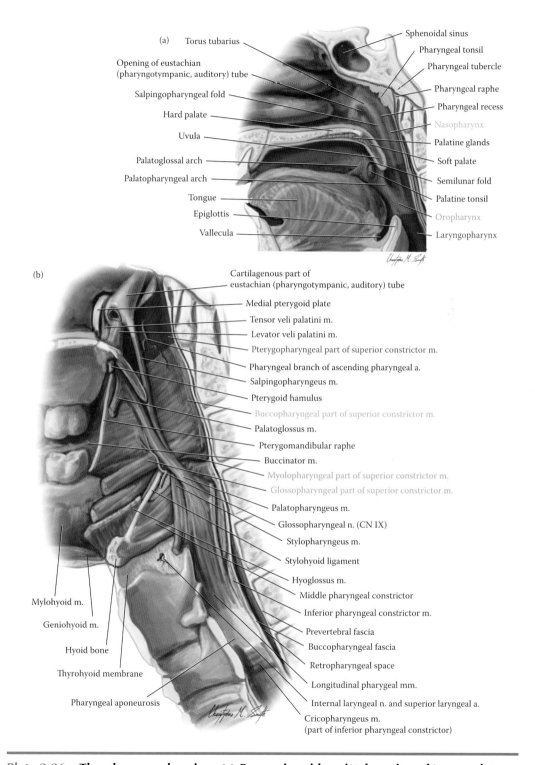

(a)

Torus tubarius

Opening of eustachian (pharyngotympanic, auditory) tube

Salpingopharyngeal fold

Hard palate

Uvula

Palatoglossal arch

Palatopharyngeal arch

Tongue

Epiglottis

Vallecula

Sphenoidal sinus

Pharyngeal tonsil

Pharyngeal tubercle

Pharyngeal raphe

Pharyngeal recess

Nasopharynx

Palatine glands

Soft palate

Semilunar fold

Palatine tonsil

Oropharynx

Laryngopharynx

(b)

Cartilagenous part of eustachian (pharyngotympanic, auditory) tube

Medial pterygoid plate

Tensor veli palatini m.

Levator veli palatini m.

Pterygopharyngeal part of superior constrictor m.

Pharyngeal branch of ascending pharyngeal a.

Salpingopharyngeus m.

Pterygoid hamulus

Buccopharyngeal part of superior constrictor m.

Palatoglossus m.

Pterygomandibular raphe

Buccinator m.

Myolopharyngeal part of superior constrictor m.

Glossopharyngeal part of superior constrictor m.

Palatopharyngeus m.

Glossopharyngeal n. (CN IX)

Stylopharyngeus m.

Stylohyoid ligament

Hyoglossus m.

Middle pharyngeal constrictor

Inferior pharyngeal constrictor m.

Prevertebral fascia

Buccopharyngeal fascia

Retropharyngeal space

Longitudinal pharygeal mm.

Internal laryngeal n. and superior laryngeal a.

Cricopharyngeus m. (part of inferior pharyngeal constrictor)

Mylohyoid m.

Geniohyoid m.

Hyoid bone

Thyrohyoid membrane

Pharyngeal aponeurosis

Plate 3.36   The pharyngeal region. (a) Fauces in mid-sagittal section; (b) musculature of the pharynx in sagittal section with mucosa removed.

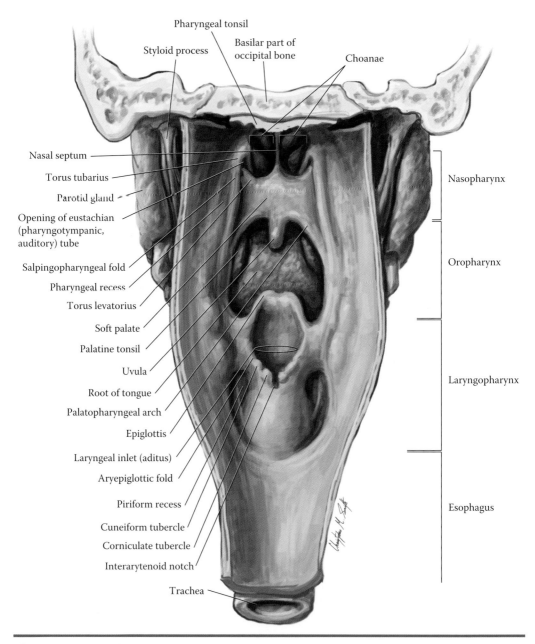

Pharyngeal tonsil

Styloid process

Basilar part of occipital bone

Choanae

Nasal septum

Torus tubarius

Parotid gland

Opening of eustachian (pharyngotympanic, auditory) tube

Salpingopharyngeal fold

Pharyngeal recess

Torus levatorius

Soft palate

Palatine tonsil

Uvula

Root of tongue

Palatopharyngeal arch

Epiglottis

Laryngeal inlet (aditus)

Aryepiglottic fold

Piriform recess

Cuneiform tubercle

Corniculate tubercle

Interarytenoid notch

Trachea

Nasopharynx

Oropharynx

Laryngopharynx

Esophagus

Plate 3.37 **The pharynx, opened in posterior view.**

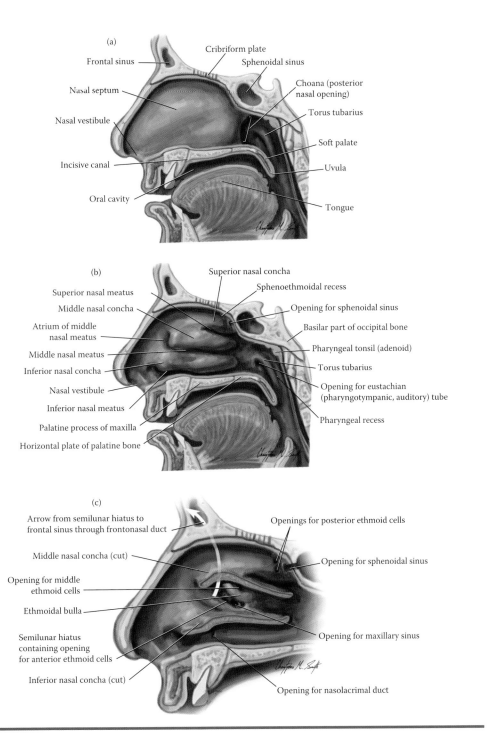

Plate 3.38   The nasal cavity in mid-sagittal section showing nasal mucosa. (a) Nasal septum intact; (b) nasal septum removed to show nasal conchae; (c) detail of lateral nasal wall with nasal conchae cut to show openings that communicate with other parts of the head.

(a)

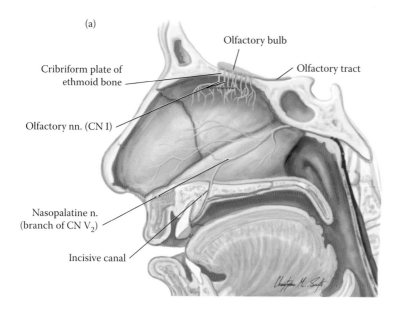

Olfactory bulb

Cribriform plate of ethmoid bone

Olfactory tract

Olfactory nn. (CN I)

Nasopalatine n. (branch of CN V₂)

Incisive canal

(b)

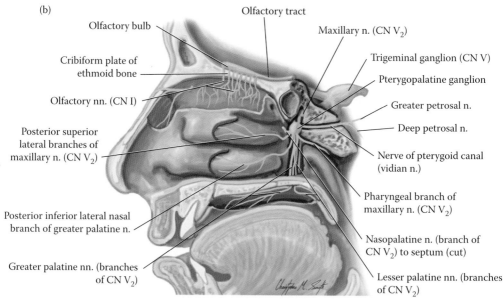

Olfactory tract

Olfactory bulb

Maxillary n. (CN V₂)

Cribiform plate of ethmoid bone

Trigeminal ganglion (CN V)

Pterygopalatine ganglion

Olfactory nn. (CN I)

Greater petrosal n.

Posterior superior lateral branches of maxillary n. (CN V₂)

Deep petrosal n.

Nerve of pterygoid canal (vidian n.)

Pharyngeal branch of maxillary n. (CN V₂)

Posterior inferior lateral nasal branch of greater palatine n.

Nasopalatine n. (branch of CN V₂) to septum (cut)

Greater palatine nn. (branches of CN V₂)

Lesser palatine nn. (branches of CN V₂)

Plate 3.39 **Innervation of the nasal region, mid-sagittal section. (a) Nasal septum intact; (b) nasal septum and parts of mucosa removed to show innervation of the lateral nasal wall and palate.**

(a)

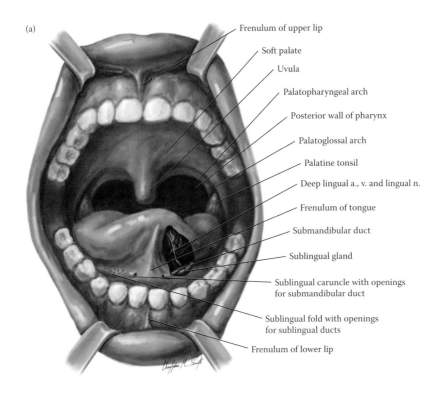

Frenulum of upper lip

Soft palate

Uvula

Palatopharyngeal arch

Posterior wall of pharynx

Palatoglossal arch

Palatine tonsil

Deep lingual a., v. and lingual n.

Frenulum of tongue

Submandibular duct

Sublingual gland

Sublingual caruncle with openings
for submandibular duct

Sublingual fold with openings
for sublingual ducts

Frenulum of lower lip

(b)

Epiglottis

Median glossoepiglottic fold

Lateral glossoepiglottic fold

Palatopharyngeal arch

Vallecula

Palatine tonsil

Lingual tonsil (lingual
nodules)

Palatoglossal arch

Foramen cecum

Terminal sulcus

Foliate papillae

Vallate papillae

Median sulcus

Filiform papillae

Fungiform papillae

Apex

Root

Body

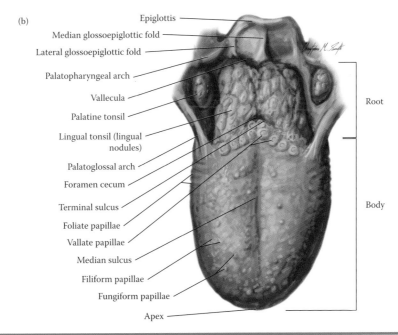

Plate 3.40   The oral cavity. (a) Anterior view of oral region with a portion of tongue epithelium removed to show deep lingual nerve and vessels; (b) superior view of tongue.

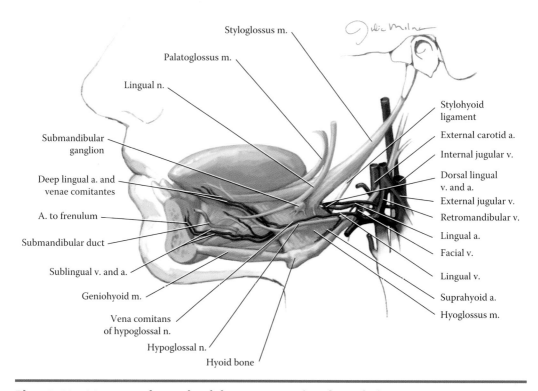

Plate 3.41   Nerves and vessels of the tongue region, lateral view.

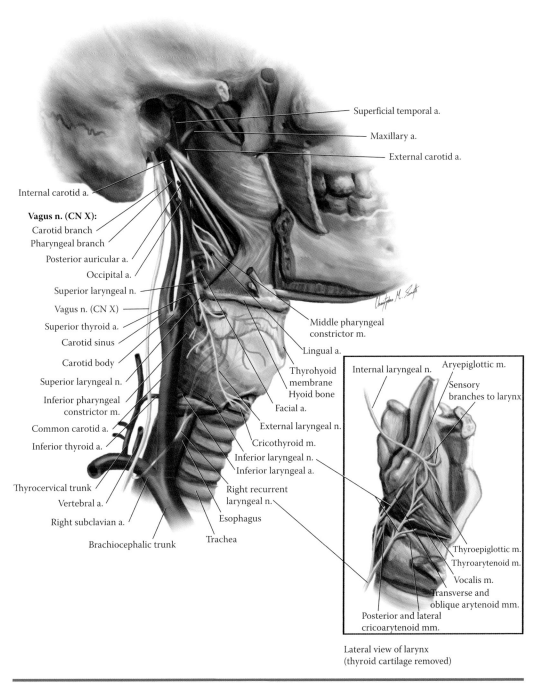

Superficial temporal a.

Maxillary a.

External carotid a.

Internal carotid a.

Vagus n. (CN X):
Carotid branch
Pharyngeal branch
Posterior auricular a.
Occipital a.
Superior laryngeal n.
Vagus n. (CN X)
Superior thyroid a.
Carotid sinus
Carotid body
Superior laryngeal n.
Inferior pharyngeal
constrictor m.
Common carotid a.
Inferior thyroid a.
Thyrocervical trunk
Vertebral a.
Right subclavian a.
Brachiocephalic trunk

Middle pharyngeal
constrictor m.
Lingual a.
Thyrohyoid
membrane
Hyoid bone
Facial a.
External laryngeal n.
Cricothyroid m.
Inferior laryngeal n.
Inferior laryngeal a.
Right recurrent
laryngeal n.
Esophagus
Trachea

Internal laryngeal n.
Aryepiglottic m.
Sensory
branches to larynx
Thyroepiglottic m.
Thyroarytenoid m.
Vocalis m.
Transverse and
oblique arytenoid mm.
Posterior and lateral
cricoarytenoid mm.

Lateral view of larynx
(thyroid cartilage removed)

Plate 3.42   Arteries and nerves of the larynx and pharynx, right lateral view. Inset: Sagittal section of the larynx showing internal innervation of the larynx.

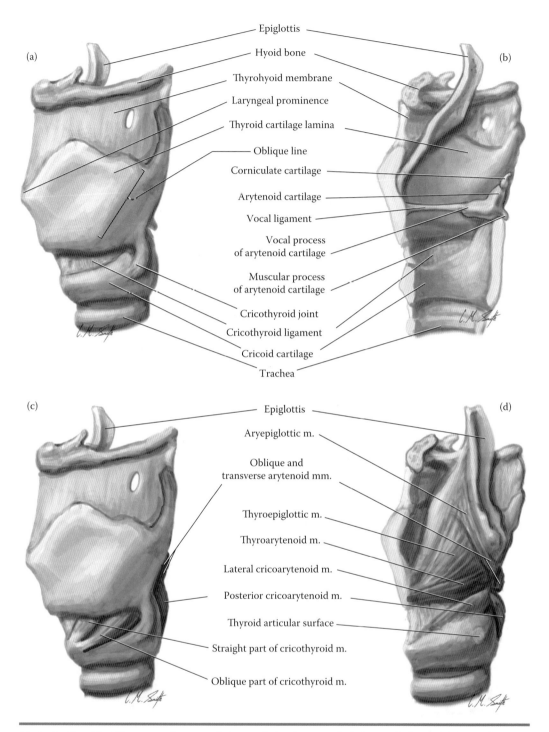

(a)

Epiglottis

Hyoid bone

Thyrohyoid membrane

Laryngeal prominence

Thyroid cartilage lamina

Oblique line

Corniculate cartilage

Arytenoid cartilage

Vocal ligament

Vocal process
of arytenoid cartilage

Muscular process
of arytenoid cartilage

Cricothyroid joint

Cricothyroid ligament

Cricoid cartilage

Trachea

(b)

(c)

Epiglottis

Aryepiglottic m.

Oblique and
transverse arytenoid mm.

Thyroepiglottic m.

Thyroarytenoid m.

Lateral cricoarytenoid m.

Posterior cricoarytenoid m.

Thyroid articular surface

Straight part of cricothyroid m.

Oblique part of cricothyroid m.

(d)

Plate 3.43 **Cartilages and musculature of the larynx, left lateral view. (a) Muscles removed; (b) mid-sagittal section with muscles removed; (c) muscles intact; (d) part of hyoid bone and thyroid cartilage removed to show internal laryngeal musculature.**

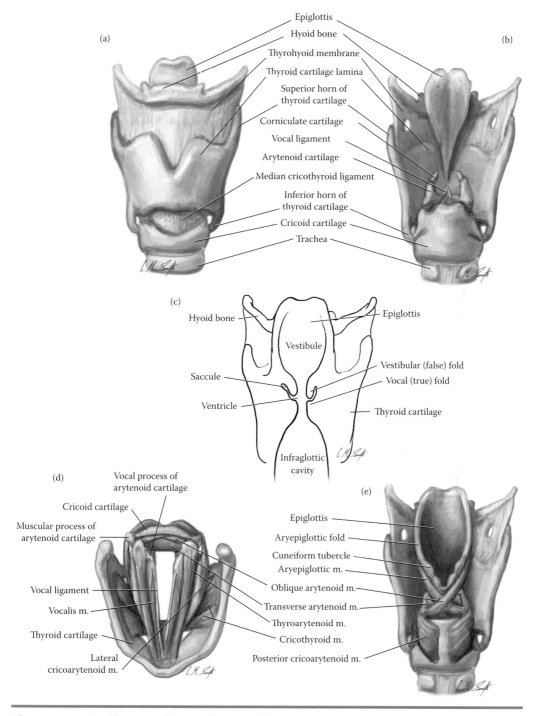

Plate 3.44 **Cartilages and musculature of the larynx. (a) Anterior view with mucosa and muscles removed; (b) posterior view with mucosa and muscles removed; (c) coronal section showing air spaces; (d) superior view; (e) posterior view.**

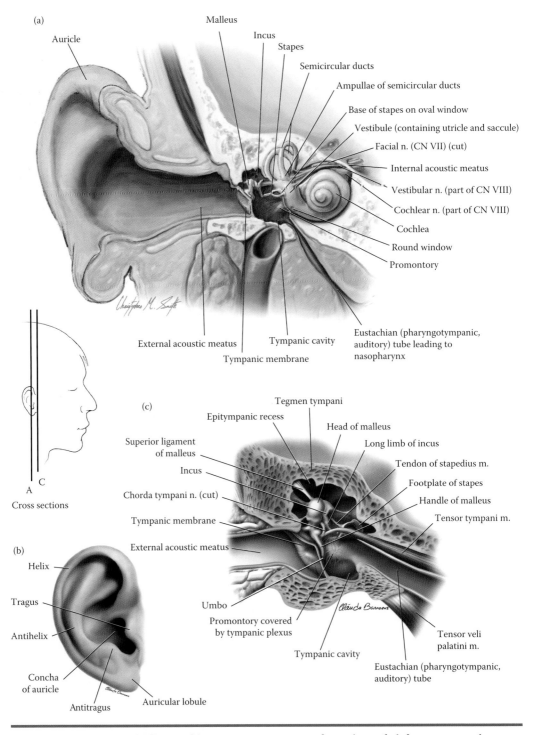

(a)

Auricle

Malleus

Incus

Stapes

Semicircular ducts

Ampullae of semicircular ducts

Base of stapes on oval window

Vestibule (containing utricle and saccule)

Facial n. (CN VII) (cut)

Internal acoustic meatus

Vestibular n. (part of CN VIII)

Cochlear n. (part of CN VIII)

Cochlea

Round window

Promontory

External acoustic meatus

Tympanic cavity

Tympanic membrane

Eustachian (pharyngotympanic, auditory) tube leading to nasopharynx

A   C

Cross sections

(c)

Tegmen tympani

Epitympanic recess

Head of malleus

Superior ligament of malleus

Long limb of incus

Tendon of stapedius m.

Incus

Footplate of stapes

Chorda tympani n. (cut)

Handle of malleus

Tympanic membrane

Tensor tympani m.

External acoustic meatus

(b)

Helix

Tragus

Antihelix

Umbo

Promontory covered by tympanic plexus

Tympanic cavity

Tensor veli palatini m.

Eustachian (pharyngotympanic, auditory) tube

Concha of auricle

Antitragus

Auricular lobule

Plate 3.45   **Outer, middle, and inner ear. (a) Coronal section of right ear, anterior view; (b) external ear; (c) detail of coronal section of right middle ear, anterior view.**

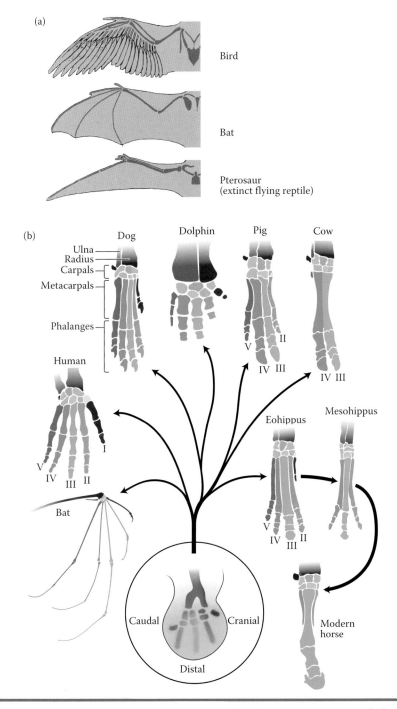

Plate 4.1 **Limb diversity. (a) Forelimb (upper limb in human anatomy) skeletons in a bird, bat, and pterosaur (extinct flying reptile); note that all have a similar number of intrinsic wing skeletal elements, but their proportions vary greatly; (b) outline of forelimb variations among mammals; underlying this diversity is a highly conserved plan for forelimb development (shown in the circle).**

**Forelimb development
(human)**

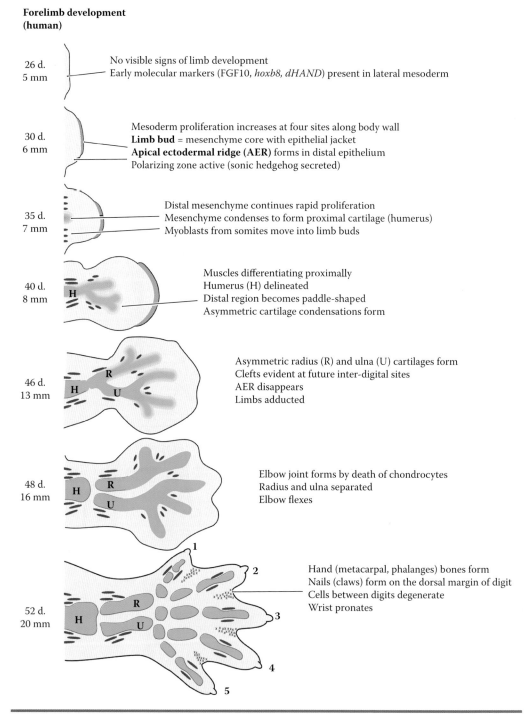

26 d.
5 mm

No visible signs of limb development
Early molecular markers (FGF10, *hoxb8, dHAND*) present in lateral mesoderm

30 d.
6 mm

Mesoderm proliferation increases at four sites along body wall
Limb bud = mesenchyme core with epithelial jacket
Apical ectodermal ridge (AER) forms in distal epithelium
Polarizing zone active (sonic hedgehog secreted)

35 d.
7 mm

Distal mesenchyme continues rapid proliferation
Mesenchyme condenses to form proximal cartilage (humerus)
Myoblasts from somites move into limb buds

40 d.
8 mm

Muscles differentiating proximally
Humerus (H) delineated
Distal region becomes paddle-shaped
Asymmetric cartilage condensations form

46 d.
13 mm

Asymmetric radius (R) and ulna (U) cartilages form
Clefts evident at future inter-digital sites
AER disappears
Limbs adducted

48 d.
16 mm

Elbow joint forms by death of chondrocytes
Radius and ulna separated
Elbow flexes

52 d.
20 mm

Hand (metacarpal, phalanges) bones form
Nails (claws) form on the dorsal margin of digit
Cells between digits degenerate
Wrist pronates

**Plate 4.2 Stages in mammalian forelimb (upper limb in human anatomy)
development.**

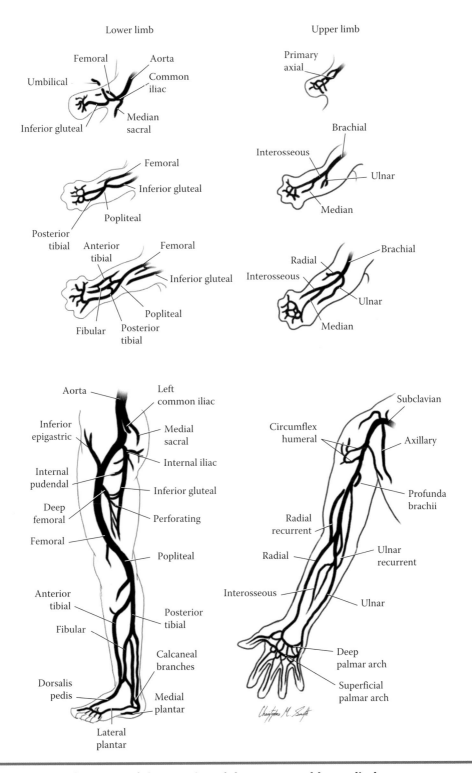

Plate 4.3   Development of the arteries of the upper and lower limbs.

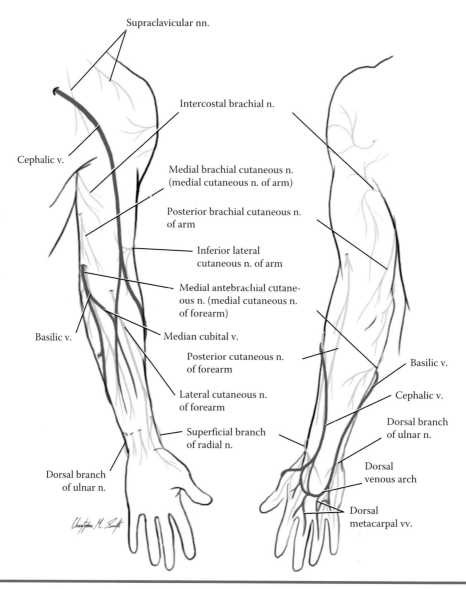

Supraclavicular nn.

Intercostal brachial n.

Cephalic v.

Medial brachial cutaneous n.
(medial cutaneous n. of arm)

Posterior brachial cutaneous n.
of arm

Inferior lateral
cutaneous n. of arm

Medial antebrachial cutaneous n. (medial cutaneous n.
of forearm)

Basilic v.

Median cubital v.

Posterior cutaneous n.
of forearm

Basilic v.

Lateral cutaneous n.
of forearm

Cephalic v.

Superficial branch
of radial n.

Dorsal branch
of ulnar n.

Dorsal branch
of ulnar n.

Dorsal
venous arch

Dorsal
metacarpal vv.

Plate 4.4   Diagram of superficial veins and nerves of the upper limb, anterior and posterior views.

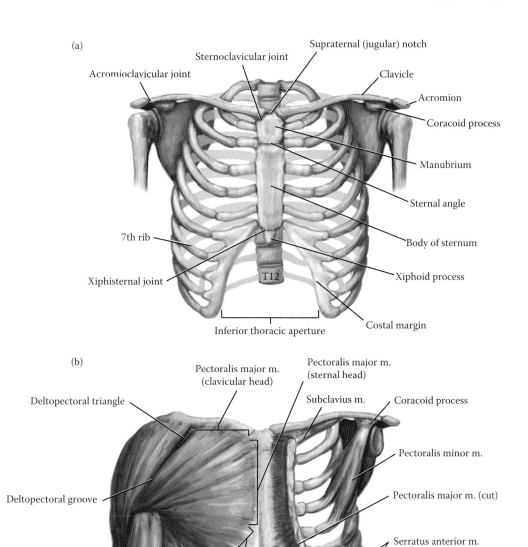

Plate 4.5 **Bones and musculature of the thorax. (a) Thoracic skeleton, anterior view; (b) pectoral musculature, anterior view. Colors correspond to developmental groups of muscles: purple—deltoid (and teres minor, not shown); red—pectoralis major and minor; blue—subclavius; green—serratus anterior.**

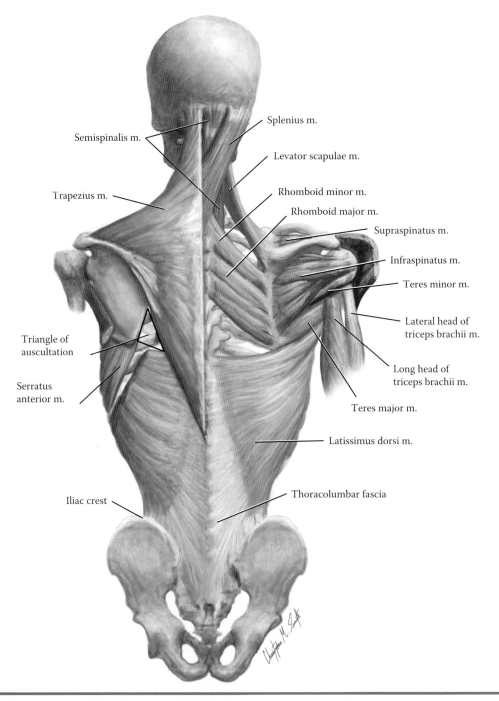

Semispinalis m.

Splenius m.

Levator scapulae m.

Trapezius m.

Rhomboid minor m.

Rhomboid major m.

Supraspinatus m.

Infraspinatus m.

Teres minor m.

Lateral head of triceps brachii m.

Triangle of auscultation

Long head of triceps brachii m.

Serratus anterior m.

Teres major m.

Latissimus dorsi m.

Thoracolumbar fascia

Iliac crest

Plate 4.6 **Superficial back musculature, posterior view. Colors correspond to developmental groups of muscles: orange—rhomboids and levator scapulae; pink—supraspinatus and infraspinatus; purple—deltoid and teres minor; light blue—latissimus dorsi and teres major (and subscapularis, not shown); green—serratus anterior.**

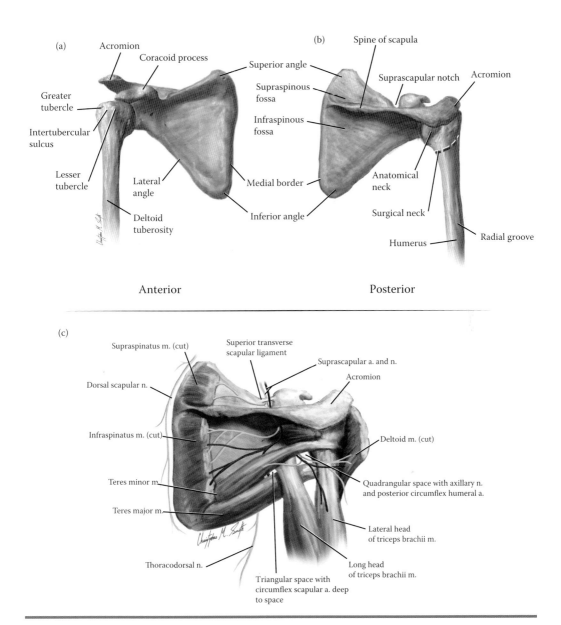

(a) Acromion
Coracoid process
Greater tubercle
Intertubercular sulcus
Lesser tubercle
Lateral angle
Deltoid tuberosity

(b) Spine of scapula
Superior angle
Suprascapular notch
Acromion
Supraspinous fossa
Infraspinous fossa
Medial border
Anatomical neck
Inferior angle
Surgical neck
Humerus
Radial groove

Anterior   Posterior

(c)
Supraspinatus m. (cut)
Superior transverse scapular ligament
Suprascapular a. and n.
Acromion
Dorsal scapular n.
Infraspinatus m. (cut)
Deltoid m. (cut)
Teres minor m.
Quadrangular space with axillary n. and posterior circumflex humeral a.
Teres major m.
Lateral head of triceps brachii m.
Thoracodorsal n.
Long head of triceps brachii m.
Triangular space with circumflex scapular a. deep to space

Plate 4.7   **The shoulder. (a) Scapula and glenohumeral joint, anterior view; (b) scapula and glenohumeral joint, posterior view; (c) the shoulder region, posterior view.**

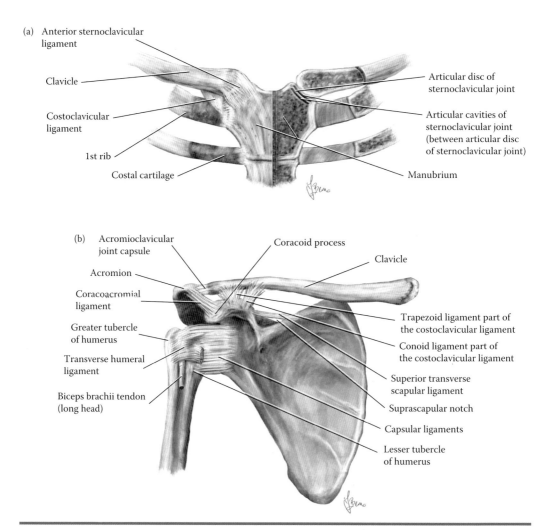

(a) Anterior sternoclavicular ligament

Clavicle

Costoclavicular ligament

1st rib

Costal cartilage

Articular disc of sternoclavicular joint

Articular cavities of sternoclavicular joint (between articular disc of sternoclavicular joint)

Manubrium

(b) Acromioclavicular joint capsule

Acromion

Coracoacromial ligament

Greater tubercle of humerus

Transverse humeral ligament

Biceps brachii tendon (long head)

Coracoid process

Clavicle

Trapezoid ligament part of the costoclavicular ligament

Conoid ligament part of the costoclavicular ligament

Superior transverse scapular ligament

Suprascapular notch

Capsular ligaments

Lesser tubercle of humerus

Plate 4.8   Joints of the shoulder girdle, anterior view. (a) Sternoclavicular joint with part of the manubrium and clavicle removed on the left side to show the joint cavity; (b) acromioclavicular and glenohumeral joints.

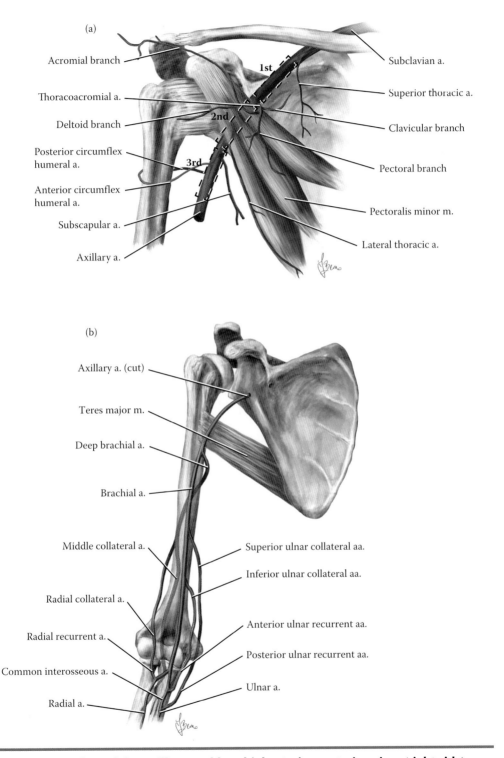

Plate 4.9   Branches of the axillary and brachial arteries, anterior view (right side).
(a) Axillary artery in the shoulder; (b) brachial artery in the arm and forearm.

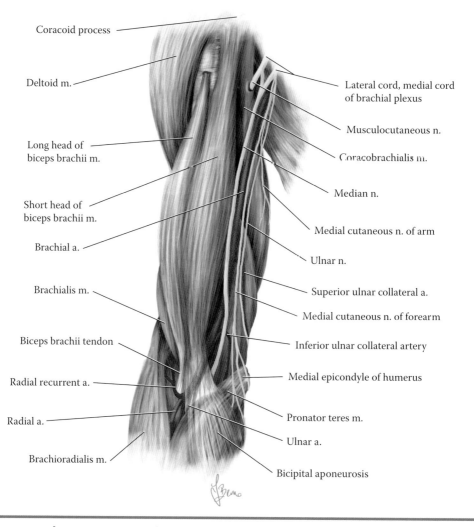

Plate 4.10   **The upper arm (right side), anterior view.**

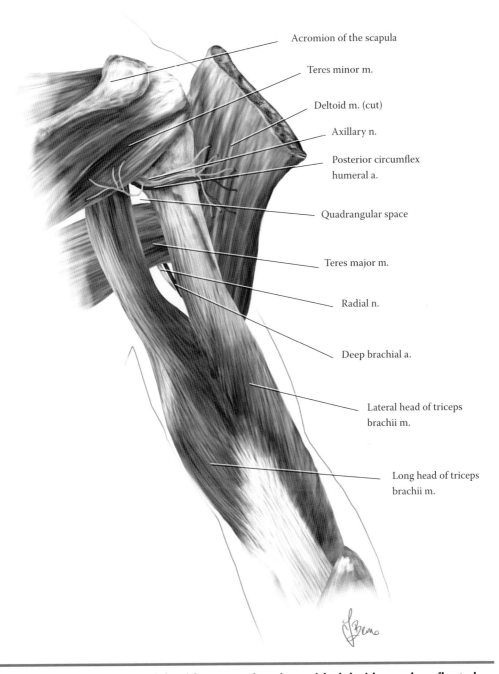

Acromion of the scapula

Teres minor m.

Deltoid m. (cut)

Axillary n.

Posterior circumflex
humeral a.

Quadrangular space

Teres major m.

Radial n.

Deep brachial a.

Lateral head of triceps
brachii m.

Long head of triceps
brachii m.

Plate 4.11   The upper arm (right side), posterior view with deltoid muscle reflected.

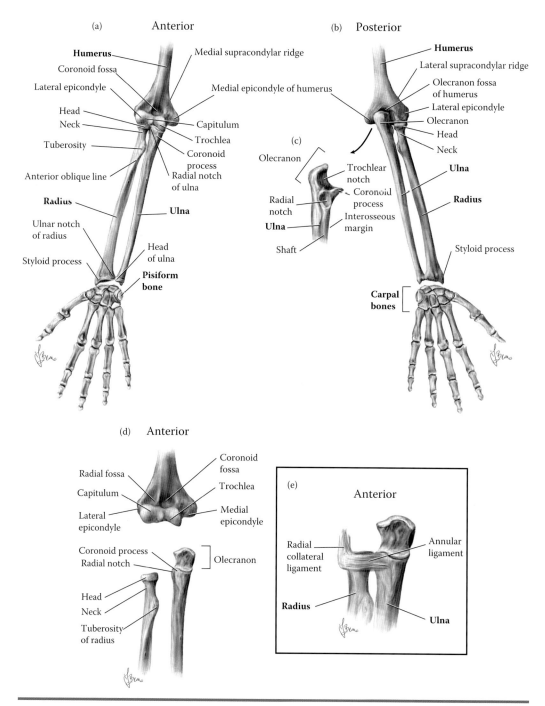

(a) Anterior

Humerus
Coronoid fossa
Lateral epicondyle
Head
Neck
Tuberosity
Anterior oblique line
Radius
Ulnar notch of radius
Styloid process

Medial supracondylar ridge
Medial epicondyle of humerus
Capitulum
Trochlea
Coronoid process
Radial notch of ulna
Ulna
Head of ulna
Pisiform bone

(b) Posterior

Humerus
Lateral supracondylar ridge
Olecranon fossa of humerus
Lateral epicondyle
Olecranon
Head
Neck
Ulna
Radius
Styloid process
Carpal bones

(c)

Olecranon
Radial notch
Ulna
Shaft

Trochlear notch
Coronoid process
Interosseous margin

(d) Anterior

Radial fossa
Capitulum
Lateral epicondyle

Coronoid fossa
Trochlea
Medial epicondyle

Coronoid process
Radial notch
Head
Neck
Tuberosity of radius

Olecranon

(e) Anterior

Radial collateral ligament
Radius

Annular ligament
Ulna

Plate 4.12   Bones of the elbow and forearm (left side). (a) Anterior view; (b) posterior view; (c) detail of proximal ulna in lateral view showing trochlear notch; (d) disarticulated elbow joint in anterior view; (e) radioulnar joint in anterior view.

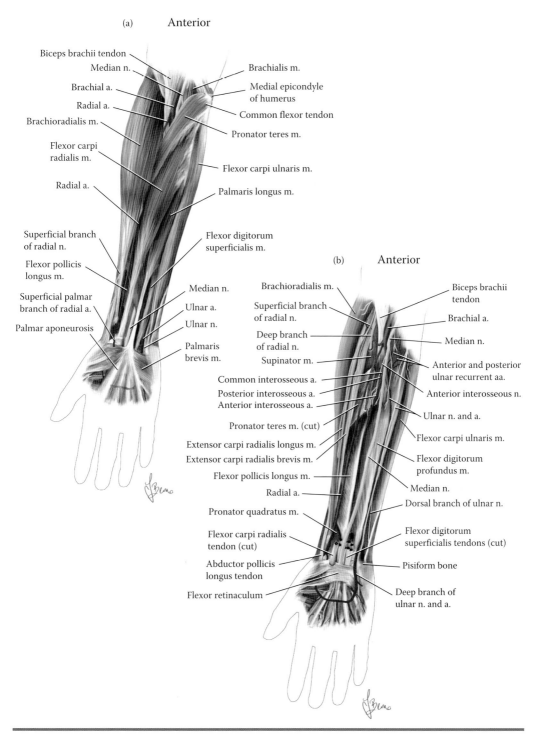

(a) Anterior

Biceps brachii tendon
Median n.
Brachial a.
Radial a.
Brachioradialis m.
Flexor carpi radialis m.
Radial a.
Superficial branch of radial n.
Flexor pollicis longus m.
Superficial palmar branch of radial a.
Palmar aponeurosis

Brachialis m.
Medial epicondyle of humerus
Common flexor tendon
Pronator teres m.
Flexor carpi ulnaris m.
Palmaris longus m.
Flexor digitorum superficialis m.

Median n.
Ulnar a.
Ulnar n.
Palmaris brevis m.

(b) Anterior

Brachioradialis m.
Superficial branch of radial n.
Deep branch of radial n.
Supinator m.
Common interosseous a.
Posterior interosseous a.
Anterior interosseous a.
Pronator teres m. (cut)
Extensor carpi radialis longus m.
Extensor carpi radialis brevis m.
Flexor pollicis longus m.
Radial a.
Pronator quadratus m.
Flexor carpi radialis tendon (cut)
Abductor pollicis longus tendon
Flexor retinaculum

Biceps brachii tendon
Brachial a.
Median n.
Anterior and posterior ulnar recurrent aa.
Anterior interosseous n.
Ulnar n. and a.
Flexor carpi ulnaris m.
Flexor digitorum profundus m.
Median n.
Dorsal branch of ulnar n.
Flexor digitorum superficialis tendons (cut)
Pisiform bone
Deep branch of ulnar n. and a.

Plate 4.13   The forearm and wrist, anterior view (right side). (a) Superficial muscles of the anterior compartment; (b) deep structures of the anterior compartment.

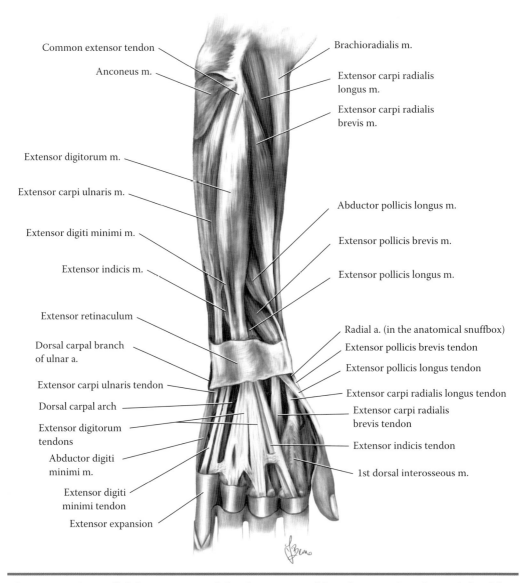

Common extensor tendon

Anconeus m.

Extensor digitorum m.

Extensor carpi ulnaris m.

Extensor digiti minimi m.

Extensor indicis m.

Extensor retinaculum

Dorsal carpal branch
of ulnar a.

Extensor carpi ulnaris tendon

Dorsal carpal arch

Extensor digitorum
tendons

Abductor digiti
minimi m.

Extensor digiti
minimi tendon

Extensor expansion

Brachioradialis m.

Extensor carpi radialis
longus m.

Extensor carpi radialis
brevis m.

Abductor pollicis longus m.

Extensor pollicis brevis m.

Extensor pollicis longus m.

Radial a. (in the anatomical snuffbox)

Extensor pollicis brevis tendon

Extensor pollicis longus tendon

Extensor carpi radialis longus tendon

Extensor carpi radialis
brevis tendon

Extensor indicis tendon

1st dorsal interosseous m.

Plate 4.14   **Superficial structures of the forearm and hand, posterior view (right side).**

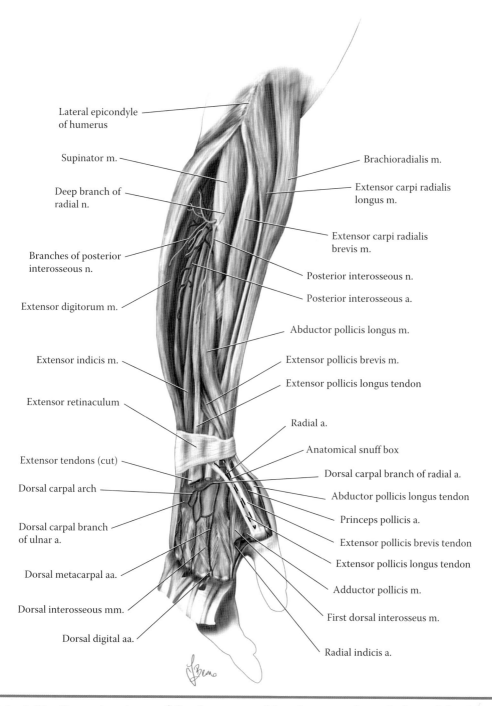

Lateral epicondyle of humerus

Supinator m.

Deep branch of radial n.

Branches of posterior interosseous n.

Extensor digitorum m.

Extensor indicis m.

Extensor retinaculum

Extensor tendons (cut)

Dorsal carpal arch

Dorsal carpal branch of ulnar a.

Dorsal metacarpal aa.

Dorsal interosseous mm.

Dorsal digital aa.

Brachioradialis m.

Extensor carpi radialis longus m.

Extensor carpi radialis brevis m.

Posterior interosseous n.

Posterior interosseous a.

Abductor pollicis longus m.

Extensor pollicis brevis m.

Extensor pollicis longus tendon

Radial a.

Anatomical snuff box

Dorsal carpal branch of radial a.

Abductor pollicis longus tendon

Princeps pollicis a.

Extensor pollicis brevis tendon

Extensor pollicis longus tendon

Adductor pollicis m.

First dorsal interosseus m.

Radial indicis a.

Plate 4.15   Deep structures of the forearm and hand, posterolateral view (right side). Extensor digitorum muscle retracted and extensor tendons cut. Dashed line indicates anatomical snuffbox.

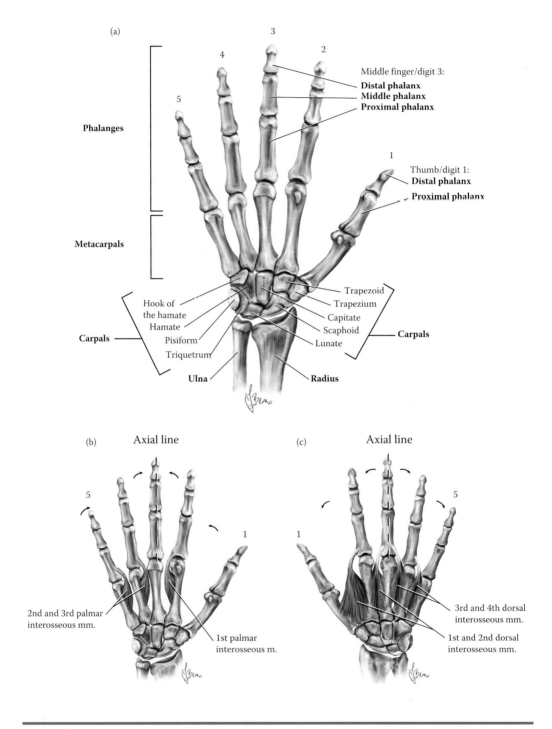

Plate 4.16 Bones of the hand (right side). (a) Anterior view; (b) palmar interosseus muscles, anterior view; (c) dorsal interosseus muscles, posterior view. Arrows show muscle actions, and dashed line indicates the axial line of the hand.

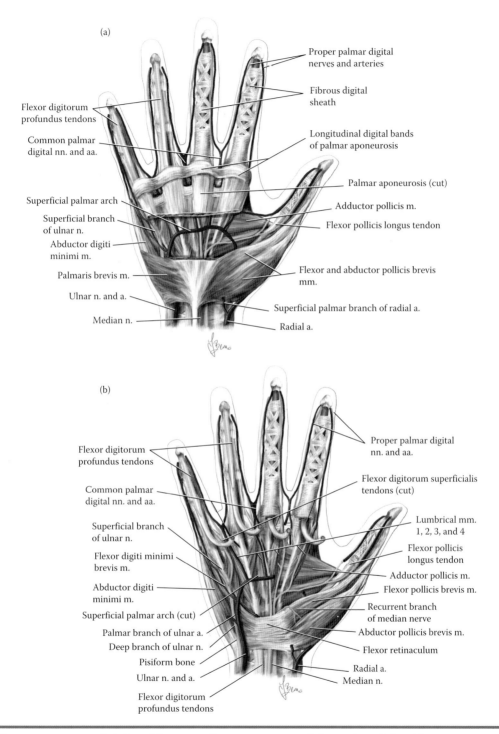

(a)

Proper palmar digital
nerves and arteries

Fibrous digital
sheath

Flexor digitorum
profundus tendons

Longitudinal digital bands
of palmar aponeurosis

Common palmar
digital nn. and aa.

Palmar aponeurosis (cut)

Superficial palmar arch

Adductor pollicis m.

Superficial branch
of ulnar n.

Flexor pollicis longus tendon

Abductor digiti
minimi m.

Palmaris brevis m.

Flexor and abductor pollicis brevis
mm.

Ulnar n. and a.

Median n.

Superficial palmar branch of radial a.

Radial a.

(b)

Proper palmar digital
nn. and aa.

Flexor digitorum
profundus tendons

Flexor digitorum superficialis
tendons (cut)

Common palmar
digital nn. and aa.

Lumbrical mm.
1, 2, 3, and 4

Superficial branch
of ulnar n.

Flexor pollicis
longus tendon

Flexor digiti minimi
brevis m.

Adductor pollicis m.

Flexor pollicis brevis m.

Abductor digiti
minimi m.

Recurrent branch
of median nerve

Superficial palmar arch (cut)

Abductor pollicis brevis m.

Palmar branch of ulnar a.

Flexor retinaculum

Deep branch of ulnar n.

Pisiform bone

Radial a.

Ulnar n. and a.

Median n.

Flexor digitorum
profundus tendons

Plate 4.17 **Superficial and intermediate structures of the palm (right side).**
(a) Superficial structures with palmar aponeurosis cut to show superficial palmar arch
arteries; (b) intermediate structures.

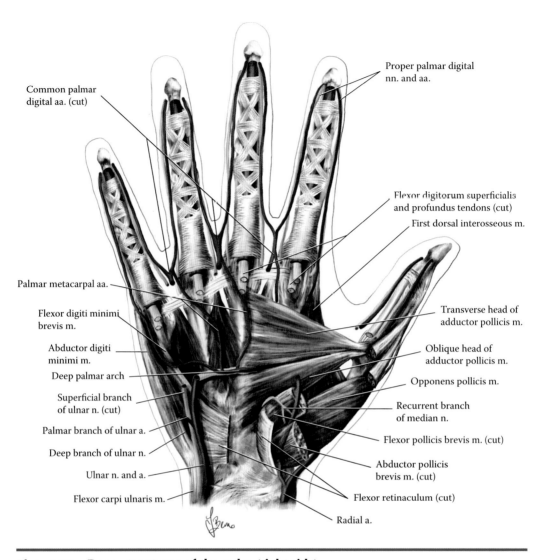

Common palmar digital aa. (cut)

Proper palmar digital nn. and aa.

Flexor digitorum superficialis and profundus tendons (cut)

First dorsal interosseous m.

Palmar metacarpal aa.

Flexor digiti minimi brevis m.

Abductor digiti minimi m.

Deep palmar arch

Superficial branch of ulnar n. (cut)

Palmar branch of ulnar a.

Deep branch of ulnar n.

Ulnar n. and a.

Flexor carpi ulnaris m.

Transverse head of adductor pollicis m.

Oblique head of adductor pollicis m.

Opponens pollicis m.

Recurrent branch of median n.

Flexor pollicis brevis m. (cut)

Abductor pollicis brevis m. (cut)

Flexor retinaculum (cut)

Radial a.

Plate 4.18   Deep structures of the palm (right side).

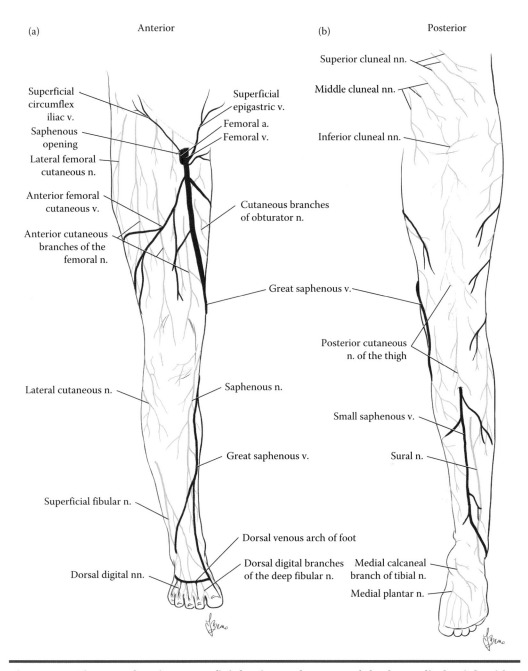

(a)   Anterior

Superficial circumflex iliac v.

Saphenous opening

Lateral femoral cutaneous n.

Anterior femoral cutaneous v.

Anterior cutaneous branches of the femoral n.

Superficial epigastric v.

Femoral a.

Femoral v.

Cutaneous branches of obturator n.

Great saphenous v.

Lateral cutaneous n.

Saphenous n.

Great saphenous v.

Superficial fibular n.

Dorsal venous arch of foot

Dorsal digital nn.

Dorsal digital branches of the deep fibular n.

(b)   Posterior

Superior cluneal nn.

Middle cluneal nn.

Inferior cluneal nn.

Great saphenous v.

Posterior cutaneous n. of the thigh

Small saphenous v.

Sural n.

Medial calcaneal branch of tibial n.

Medial plantar n.

Plate 5.1   **Diagram showing superficial veins and nerves of the lower limb (right side). (a) Anterior view; (b) posterior view.**

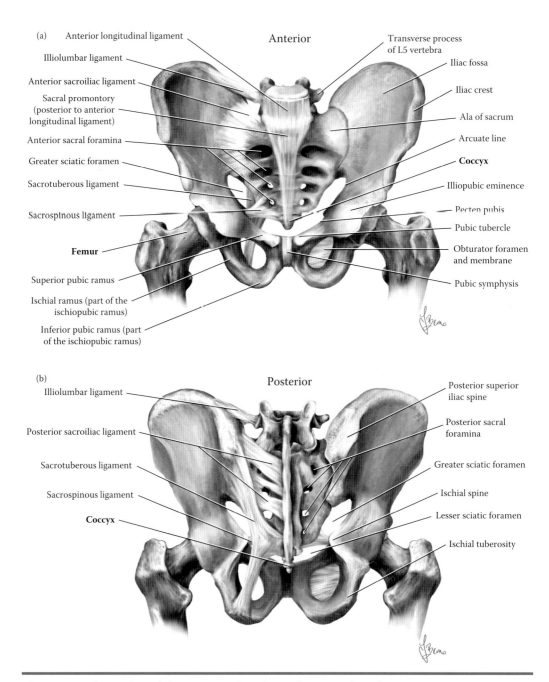

(a)

Anterior longitudinal ligament

Illiolumbar ligament

Anterior sacroiliac ligament

Sacral promontory
(posterior to anterior
longitudinal ligament)

Anterior sacral foramina

Greater sciatic foramen

Sacrotuberous ligament

Sacrospinous ligament

Femur

Superior pubic ramus

Ischial ramus (part of the
ischiopubic ramus)

Inferior pubic ramus (part
of the ischiopubic ramus)

Anterior

Transverse process
of L5 vertebra

Iliac fossa

Iliac crest

Ala of sacrum

Arcuate line

Coccyx

Illiopubic eminence

Pecten pubis

Pubic tubercle

Obturator foramen
and membrane

Pubic symphysis

(b)

Illiolumbar ligament

Posterior sacroiliac ligament

Sacrotuberous ligament

Sacrospinous ligament

Coccyx

Posterior

Posterior superior
iliac spine

Posterior sacral
foramina

Greater sciatic foramen

Ischial spine

Lesser sciatic foramen

Ischial tuberosity

Plate 5.2   **The male pelvis. (a) Anterior view; (b) posterior view.**

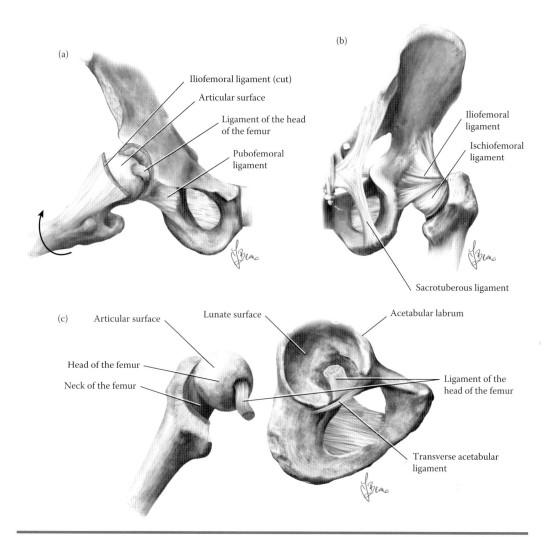

(a)

Iliofemoral ligament (cut)

Articular surface

Ligament of the head of the femur

Pubofemoral ligament

(b)

Iliofemoral ligament

Ischiofemoral ligament

Sacrotuberous ligament

(c)   Articular surface

Lunate surface

Acetabular labrum

Head of the femur

Neck of the femur

Ligament of the head of the femur

Transverse acetabular ligament

Plate 5.3   **The hip joint (right side). (a) Anterior view with iliofemoral ligament cut to show articular surface and ligament of the head of the femur; (b) posterior view; (c) disarticulated joint showing articular surfaces of the femur and acetabulum and cut ligament of the head of the femur.**

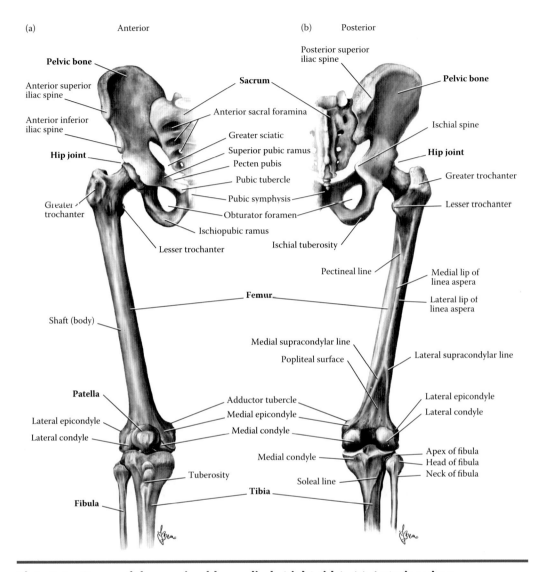

(a) Anterior

(b) Posterior

Posterior superior
iliac spine

Pelvic bone

Anterior superior
iliac spine

Sacrum

Pelvic bone

Anterior inferior
iliac spine

Anterior sacral foramina

Ischial spine

Hip joint

Greater sciatic

Superior pubic ramus

Pecten pubis

Hip joint

Greater
trochanter

Pubic tubercle

Pubic symphysis

Greater trochanter

Lesser trochanter

Obturator foramen

Ischiopubic ramus

Lesser trochanter

Ischial tuberosity

Pectineal line

Medial lip of
linea aspera

Femur

Lateral lip of
linea aspera

Shaft (body)

Medial supracondylar line

Popliteal surface

Lateral supracondylar line

Patella

Adductor tubercle

Lateral epicondyle

Medial epicondyle

Lateral epicondyle

Lateral condyle

Medial condyle

Lateral condyle

Medial condyle

Apex of fibula
Head of fibula
Neck of fibula

Tuberosity

Soleal line

Fibula

Tibia

Plate 5.4   Bones of the proximal lower limb (right side). (a) Anterior view;
(b) posterior view.

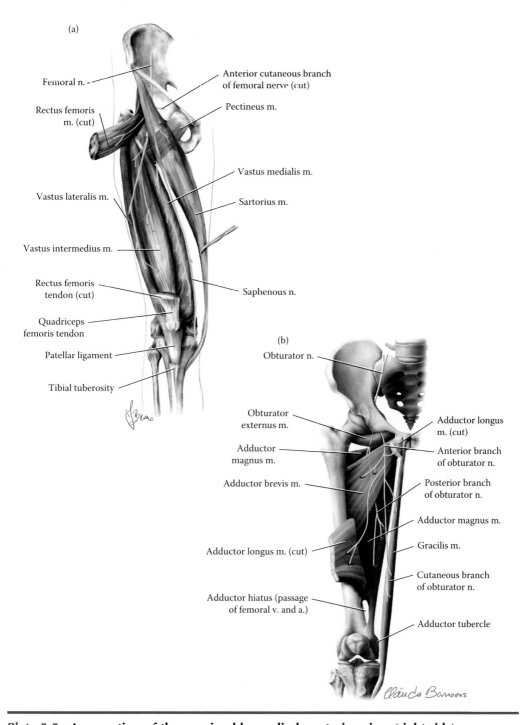

(a)

Femoral n.

Rectus femoris m. (cut)

Vastus lateralis m.

Vastus intermedius m.

Rectus femoris tendon (cut)

Quadriceps femoris tendon

Patellar ligament

Tibial tuberosity

Anterior cutaneous branch of femoral nerve (cut)

Pectineus m.

Vastus medialis m.

Sartorius m.

Saphenous n.

(b)

Obturator n.

Obturator externus m.

Adductor magnus m.

Adductor brevis m.

Adductor longus m. (cut)

Adductor hiatus (passage of femoral v. and a.)

Adductor longus m. (cut)

Anterior branch of obturator n.

Posterior branch of obturator n.

Adductor magnus m.

Gracilis m.

Cutaneous branch of obturator n.

Adductor tubercle

Plate 5.5   Innervation of the proximal lower limb, anterior view (right side).
(a) Superficial structures, with rectus femoris cut and reflected to show branches of femoral nerve; (b) deep structures, with adductor longus cut and reflected to show branches of obturator nerve.

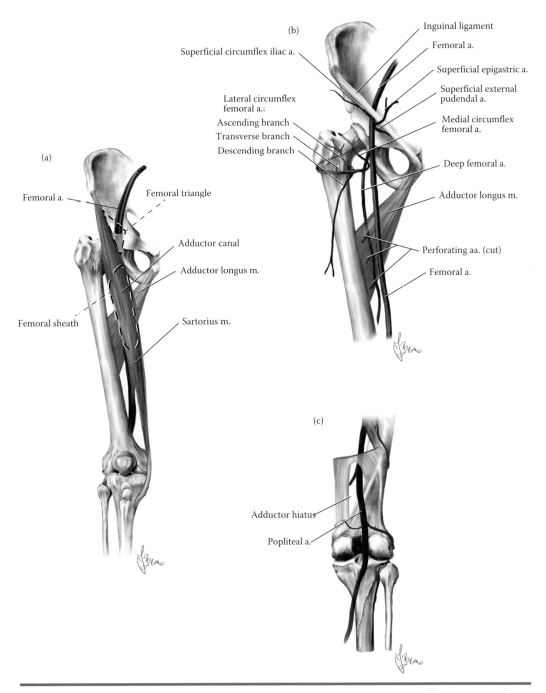

(b)

Superficial circumflex iliac a.

Inguinal ligament

Femoral a.

Superficial epigastric a.

Superficial external
pudendal a.

Lateral circumflex
femoral a.:

Ascending branch

Transverse branch

Descending branch

Medial circumflex
femoral a.

Deep femoral a.

Adductor longus m.

Perforating aa. (cut)

Femoral a.

(a)

Femoral a.

Femoral triangle

Adductor canal

Adductor longus m.

Sartorius m.

Femoral sheath

(c)

Adductor hiatus

Popliteal a.

Plate 5.6   Arteries of the proximal lower limb (right side). (a) Femoral artery passing through the adductor canal (dashed line), anterior view; (b) branches of the femoral artery, anterior view; (c) the popliteal artery emerging from the adductor hiatus (posterior view).

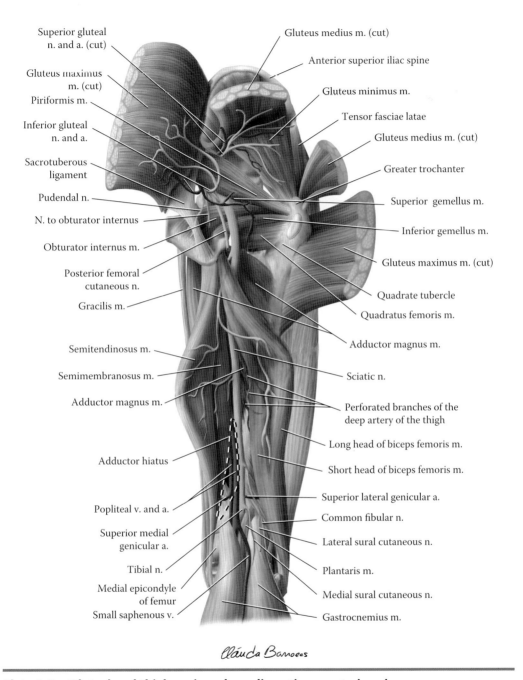

Superior gluteal n. and a. (cut)

Gluteus maximus m. (cut)

Piriformis m.

Inferior gluteal n. and a.

Sacrotuberous ligament

Pudendal n.

N. to obturator internus

Obturator internus m.

Posterior femoral cutaneous n.

Gracilis m.

Semitendinosus m.

Semimembranosus m.

Adductor magnus m.

Adductor hiatus

Popliteal v. and a.

Superior medial genicular a.

Tibial n.

Medial epicondyle of femur

Small saphenous v.

Gluteus medius m. (cut)

Anterior superior iliac spine

Gluteus minimus m.

Tensor fasciae latae

Gluteus medius m. (cut)

Greater trochanter

Superior gemellus m.

Inferior gemellus m.

Gluteus maximus m. (cut)

Quadrate tubercle

Quadratus femoris m.

Adductor magnus m.

Sciatic n.

Perforated branches of the deep artery of the thigh

Long head of biceps femoris m.

Short head of biceps femoris m.

Superior lateral genicular a.

Common fibular n.

Lateral sural cutaneous n.

Plantaris m.

Medial sural cutaneous n.

Gastrocnemius m.

Cláucia Barroes

Plate 5.7   **Gluteal and thigh region, deep dissection, posterior view.**

Anterior   Posterior

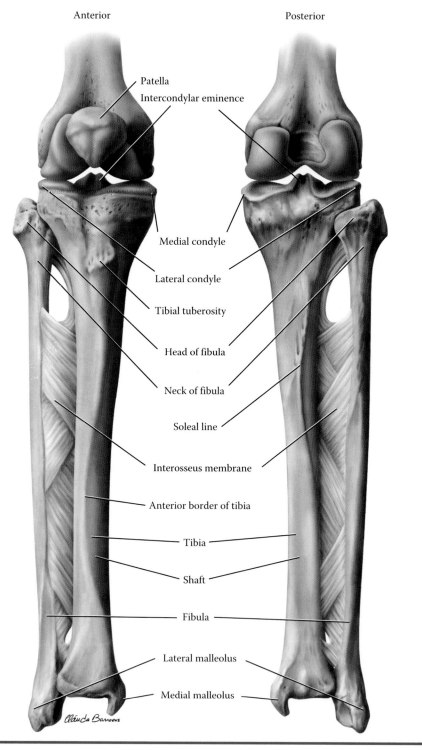

Patella

Intercondylar eminence

Medial condyle

Lateral condyle

Tibial tuberosity

Head of fibula

Neck of fibula

Soleal line

Interosseus membrane

Anterior border of tibia

Tibia

Shaft

Fibula

Lateral malleolus

Medial malleolus

Plate 5.8   **Bones of the leg.**

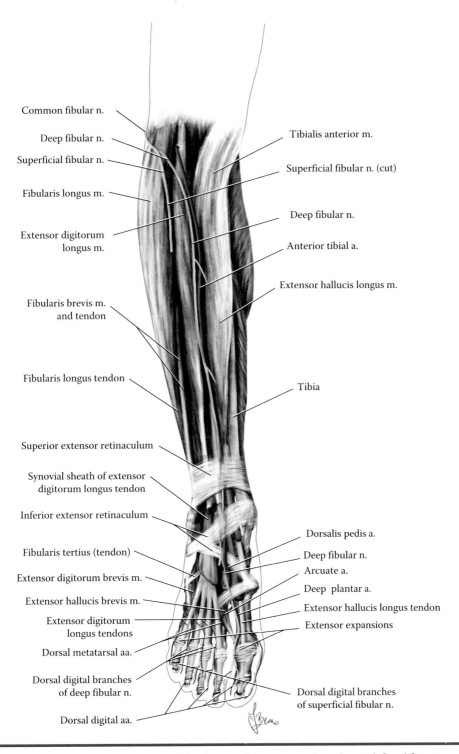

Common fibular n.

Deep fibular n.

Superficial fibular n.

Fibularis longus m.

Extensor digitorum longus m.

Fibularis brevis m. and tendon

Fibularis longus tendon

Superior extensor retinaculum

Synovial sheath of extensor digitorum longus tendon

Inferior extensor retinaculum

Fibularis tertius (tendon)

Extensor digitorum brevis m.

Extensor hallucis brevis m.

Extensor digitorum longus tendons

Dorsal metatarsal aa.

Dorsal digital branches of deep fibular n.

Dorsal digital aa.

Tibialis anterior m.

Superficial fibular n. (cut)

Deep fibular n.

Anterior tibial a.

Extensor hallucis longus m.

Tibia

Dorsalis pedis a.

Deep fibular n.

Arcuate a.

Deep plantar a.

Extensor hallucis longus tendon

Extensor expansions

Dorsal digital branches of superficial fibular n.

Plate 5.9 Superficial structures of the leg and foot, anterior view (right side). Extensor hallucis longus is retracted to show anterior tibial artery and deep fibular nerve.

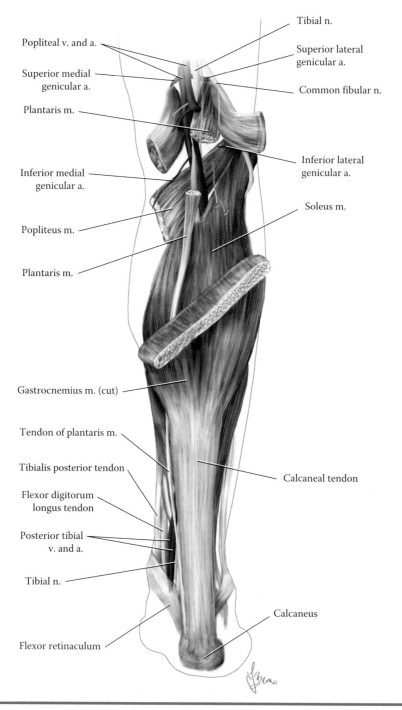

Popliteal v. and a.

Superior medial
genicular a.

Plantaris m.

Inferior medial
genicular a.

Popliteus m.

Plantaris m.

Tibial n.

Superior lateral
genicular a.

Common fibular n.

Inferior lateral
genicular a.

Soleus m.

Gastrocnemius m. (cut)

Tendon of plantaris m.

Tibialis posterior tendon

Flexor digitorum
longus tendon

Posterior tibial
v. and a.

Tibial n.

Flexor retinaculum

Calcaneal tendon

Calcaneus

**Plate 5.10   Superficial structures of the leg and foot, posterior view (right side).
Gastrocnemius is cut and reflected to show the vessels and nerves of the popliteal
region.**

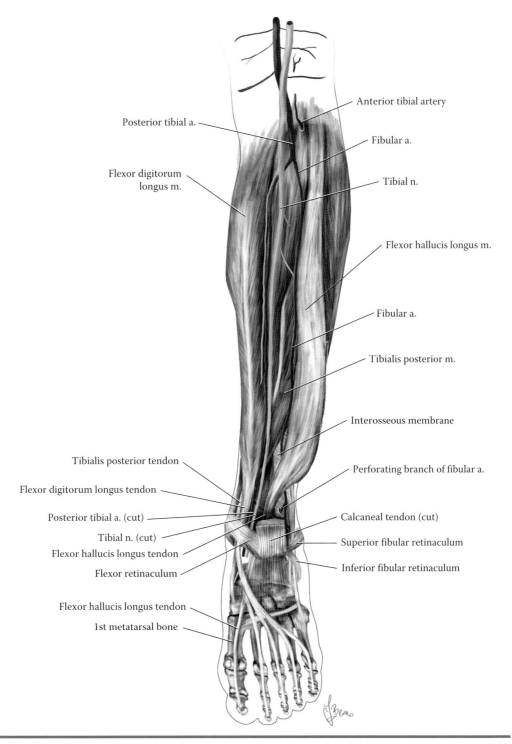

Anterior tibial artery

Posterior tibial a.

Fibular a.

Flexor digitorum longus m.

Tibial n.

Flexor hallucis longus m.

Fibular a.

Tibialis posterior m.

Interosseous membrane

Tibialis posterior tendon

Perforating branch of fibular a.

Flexor digitorum longus tendon

Posterior tibial a. (cut)

Calcaneal tendon (cut)

Tibial n. (cut)

Superior fibular retinaculum

Flexor hallucis longus tendon

Inferior fibular retinaculum

Flexor retinaculum

Flexor hallucis longus tendon

1st metatarsal bone

Plate 5.11   Deep structures of the leg and foot, posterior view (right side). Flexor hallucis longus is retracted to show the fibular artery and tibial artery and nerve.

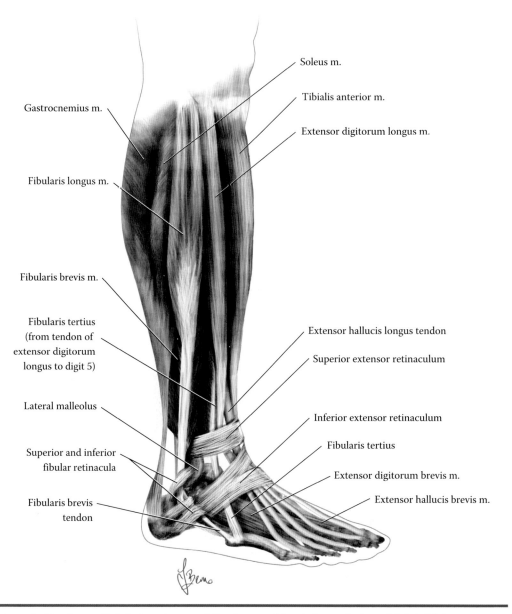

Soleus m.

Tibialis anterior m.

Extensor digitorum longus m.

Gastrocnemius m.

Fibularis longus m.

Fibularis brevis m.

Fibularis tertius
(from tendon of
extensor digitorum
longus to digit 5)

Lateral malleolus

Extensor hallucis longus tendon

Superior extensor retinaculum

Inferior extensor retinaculum

Fibularis tertius

Superior and inferior
fibular retinacula

Extensor digitorum brevis m.

Extensor hallucis brevis m.

Fibularis brevis
tendon

Plate 5.12   Superficial structures of the leg and foot, lateral view (right side).

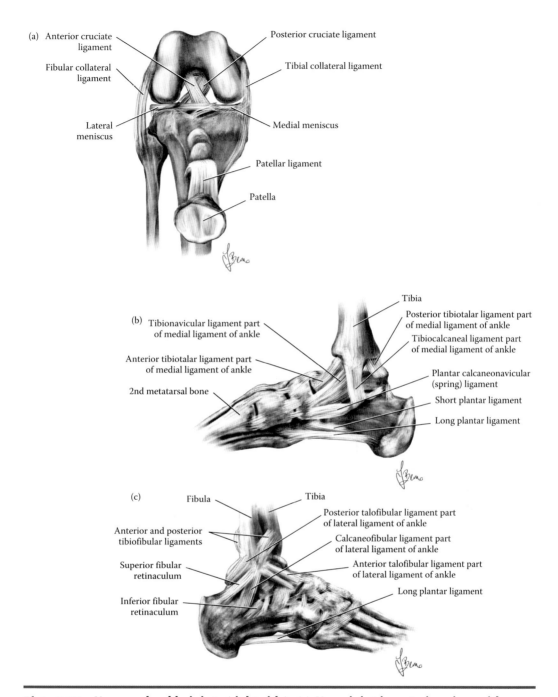

Plate 5.13   Knee and ankle joints (right side). (a) Knee joint in anterior view with patella and patellar ligament reflected to show contents of joint cavity; (b) ankle joint in medial view; (c) ankle joint in lateral view.

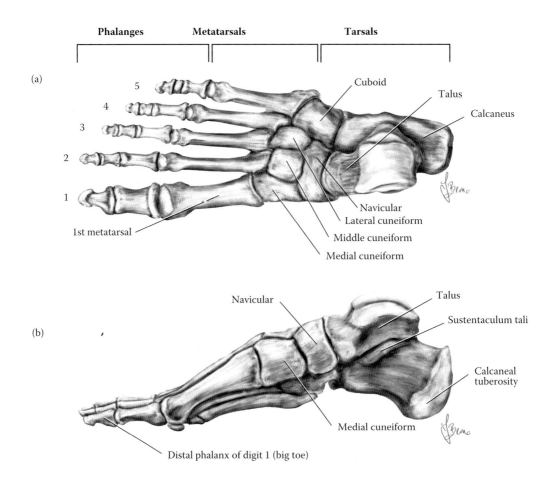

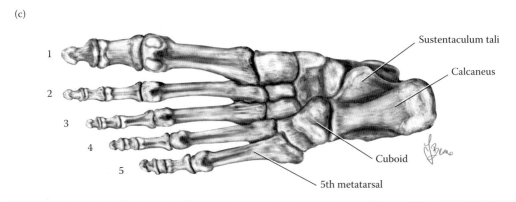

Plate 5.14   Bones of the foot (right side). (a) Dorsal (superior) view; (b) medial view; (c) plantar (inferior) view.

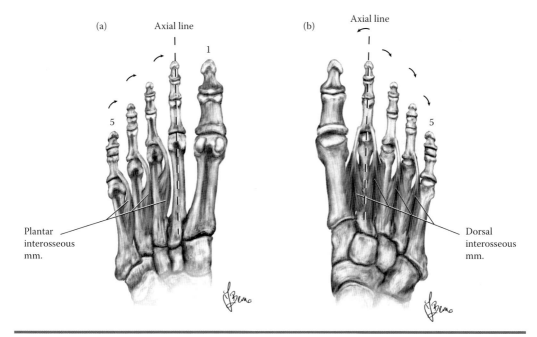

Plate 5.15   Interosseus muscles of the foot (right side). (a) Plantar interosseus muscles, plantar (inferior) view; (b) dorsal interosseus muscles, dorsal (superior) view. Arrows show muscle actions, and dashed line indicates the axial line of the foot.

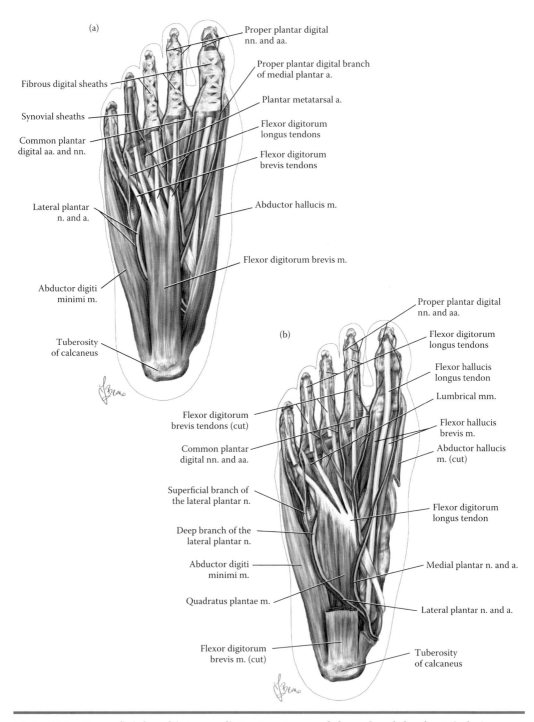

(a)

Proper plantar digital
nn. and aa.

Proper plantar digital branch
of medial plantar a.

Plantar metatarsal a.

Flexor digitorum
longus tendons

Flexor digitorum
brevis tendons

Abductor hallucis m.

Flexor digitorum brevis m.

Fibrous digital sheaths

Synovial sheaths

Common plantar
digital aa. and nn.

Lateral plantar
n. and a.

Abductor digiti
minimi m.

Tuberosity
of calcaneus

(b)

Proper plantar digital
nn. and aa.

Flexor digitorum
longus tendons

Flexor hallucis
longus tendon

Lumbrical mm.

Flexor hallucis
brevis m.

Abductor hallucis
m. (cut)

Flexor digitorum
longus tendon

Medial plantar n. and a.

Lateral plantar n. and a.

Tuberosity
of calcaneus

Flexor digitorum
brevis tendons (cut)

Common plantar
digital nn. and aa.

Superficial branch of
the lateral plantar n.

Deep branch of the
lateral plantar n.

Abductor digiti
minimi m.

Quadratus plantae m.

Flexor digitorum
brevis m. (cut)

Plate 5.16   **Superficial and intermediate structures of the sole of the foot, inferior view (right side). (a) Superficial structures/first layer of muscles; (b) intermediate structures/second layer of muscles.**

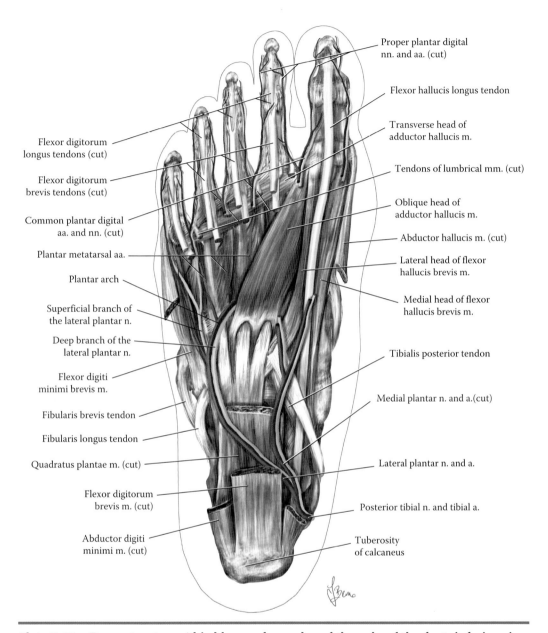

Proper plantar digital nn. and aa. (cut)

Flexor hallucis longus tendon

Transverse head of adductor hallucis m.

Tendons of lumbrical mm. (cut)

Oblique head of adductor hallucis m.

Abductor hallucis m. (cut)

Lateral head of flexor hallucis brevis m.

Medial head of flexor hallucis brevis m.

Tibialis posterior tendon

Medial plantar n. and a.(cut)

Lateral plantar n. and a.

Posterior tibial n. and tibial a.

Tuberosity of calcaneus

Flexor digitorum longus tendons (cut)

Flexor digitorum brevis tendons (cut)

Common plantar digital aa. and nn. (cut)

Plantar metatarsal aa.

Plantar arch

Superficial branch of the lateral plantar n.

Deep branch of the lateral plantar n.

Flexor digiti minimi brevis m.

Fibularis brevis tendon

Fibularis longus tendon

Quadratus plantae m. (cut)

Flexor digitorum brevis m. (cut)

Abductor digiti minimi m. (cut)

Plate 5.17 **Deep structures/third layer of muscles of the sole of the foot, inferior view (right side).**

(a)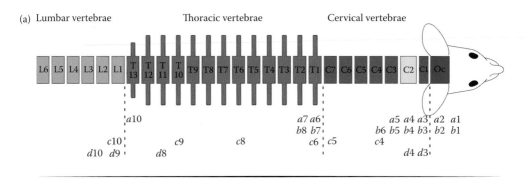

Lumbar vertebrae   Thoracic vertebrae   Cervical vertebrae

(b)

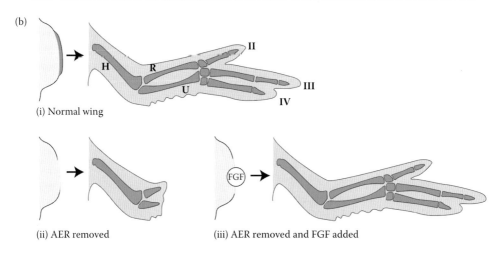

(i) Normal wing

(ii) AER removed

(iii) AER removed and FGF added

Normal limb showing zone of polarizing activity (ZPA)

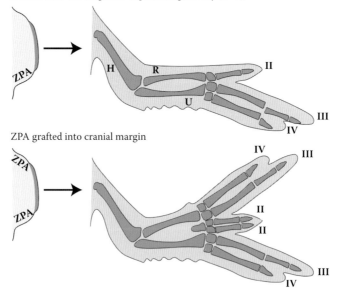

ZPA grafted into cranial margin

Plate 6.1   (*Continued*)

Plate 6.1 (Continued)   Patterning in development. (a) Cranial boundaries of *Hox* gene expression in somites (shown as vertebrae) of a mammal. Note the tendency for expression boundaries to cluster at sites of morphological transition among vertebrae. The expression of many *Hox* genes begins during gastrulation, generally sweeping from caudal to cranial within the neural plate and paraxial mesoderm. While expression can occur over an extended domain, the expression of most *Hox* genes has a precise cranial boundary, which is different for each member of a Hox cluster. (b) Effects on antero-posterior (cranio-caudal in human anatomy) asymmetry of signals from posterior limb mesoderm (ZPA), in chicken.

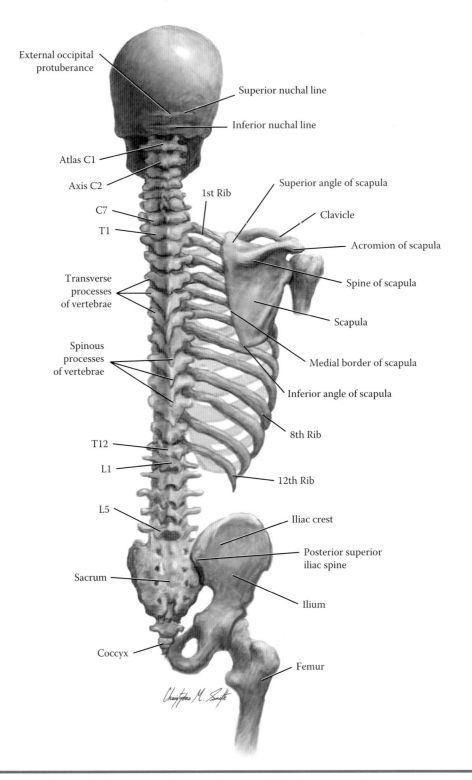

External occipital protuberance

Superior nuchal line

Inferior nuchal line

Atlas C1

Axis C2

C7

T1

1st Rib

Superior angle of scapula

Clavicle

Acromion of scapula

Spine of scapula

Transverse processes of vertebrae

Scapula

Spinous processes of vertebrae

Medial border of scapula

Inferior angle of scapula

8th Rib

T12

L1

12th Rib

L5

Iliac crest

Posterior superior iliac spine

Sacrum

Ilium

Coccyx

Femur

Plate 6.2   The axial skeleton and pectoral and pelvic girdles, posterior view.

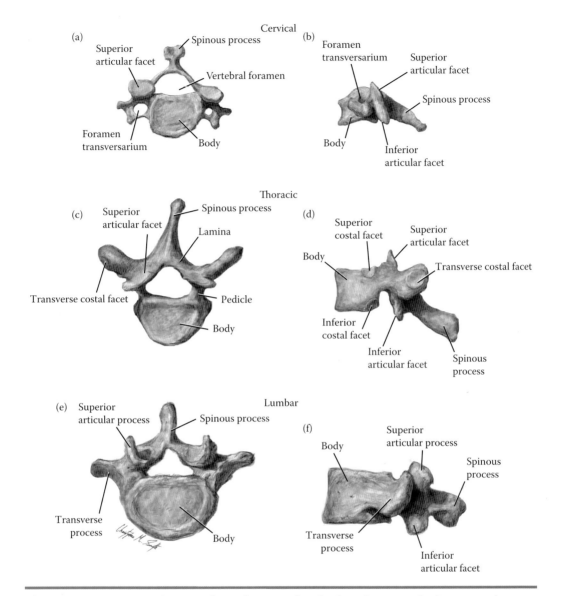

Plate 6.3 **Representative vertebrae from each spinal region: Cervical (a) superior and (b) lateral view; thoracic (c) superior and (d) lateral view; lumbar (e) superior and (f) lateral view.**

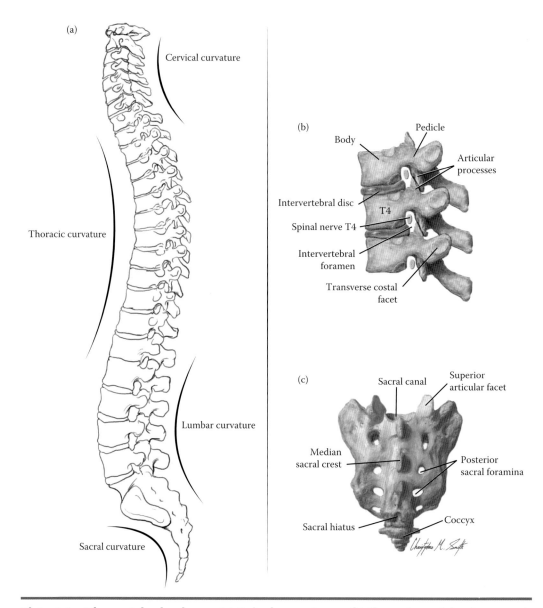

Plate 6.4   The vertebral column. (a) Spinal curvatures; (b) thoracic vertebral segment, left lateral view; (c) sacrum, posterior view.

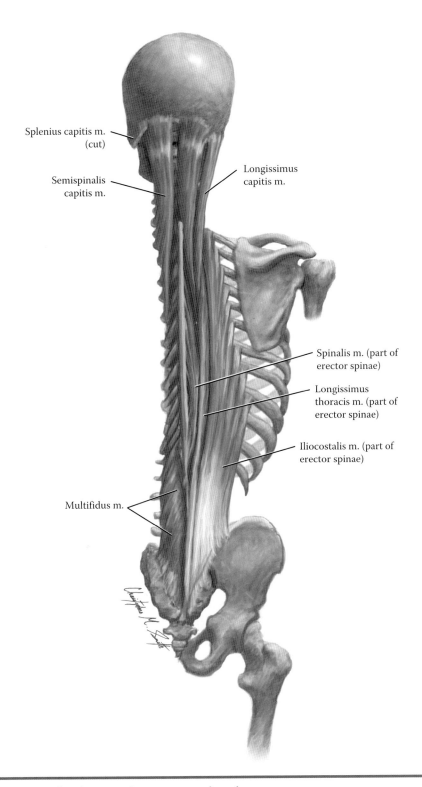

Splenius capitis m. (cut)

Semispinalis capitis m.

Longissimus capitis m.

Spinalis m. (part of erector spinae)

Longissimus thoracis m. (part of erector spinae)

Iliocostalis m. (part of erector spinae)

Multifidus m.

Plate 6.5 **Deep back musculature, posterior view.**

Index